Also by Robert A Mayers

The War Man
The True Story of a Citizen-Soldier who Fought from Quebec to Yorktown

The Portrait of an American Family
Allison-Mayers Family History

The Forgotten Revolution
When History Forgets: Revisiting Critical Places of the
American Revolution that have been Neglected by History

Searching for Private Yankee Doodle
Washington's Soldiers in the American Revolution

Revolutionary New Jersey
Forgotten Towns and Crossroads of the American Revolution

Historic Tales of Watchung

MIDDLEBROOK

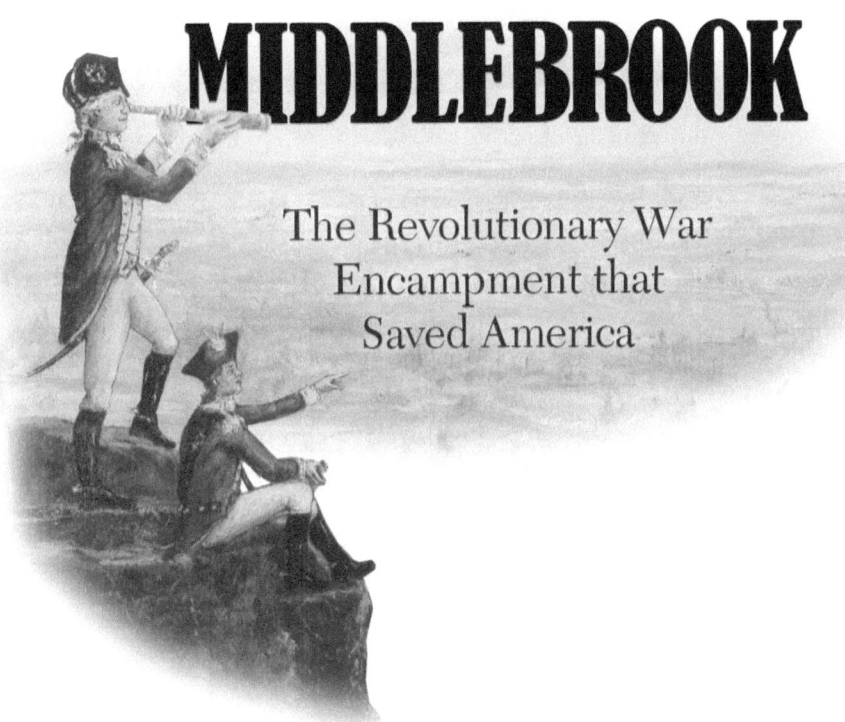

MIDDLEBROOK

The Revolutionary War Encampment that Saved America

Robert A. Mayers

American History Press
Staunton Virginia

All rights reserved. No part of this book may be transmitted in any form by any means electronic, mechanical or otherwise using devices now existing or yet to be invented without prior written permission from the publisher and copyright holder.

Staunton, Virginia
(540) 487-1202
Visit us on the Internet at:
www.Americanhistorypress.com

First Printing August 2021

To schedule an event with the author or to inquire about bulk discount sales, please contact American History Press.

Library of Congress Cataloging-in-Publication Data

Names: Mayers, Robert A. (Robert Adrian), 1930- author.
Title: Middlebrook : the Revolutionary War campground that saved America / Robert A. Mayers.
Other titles: Revolutionary War campground that saved America
Description: Staunton, Virginia : American History Press, 2021. | Includes bibliographical references and index.
Identifiers: LCCN 2021000538 | ISBN 9781939995384 (paperback)
Subjects: LCSH: Middlebrook Encampment (N.J.) | New Jersey--History--Revolution, 1775-1783--Campaigns. | Somerset County (N.J.)--History--Revolution, 1775-1783.
Classification: LCC E263.N5 M388 2021 | DDC 974.9/03--dc23
LC record available at https://lccn.loc.gov/2021000538

Manufactured in the United States of America on acid-free paper.
This book exceeds all ANSO standards for archival quality.

DEDICATION

For Herb Patullo and Kels Swan, fondly remembered and often in my thoughts. They devoted their lives to keeping the spirit of the Middlebrook Campground alive. And for Mayor Bob Fazen of Bound Brook, President of the Washington Campground Association, and the many local historians who joined me in exploring this fascinating place. I salute their eagerness to share treasured details not found in history books, their reverence for this cherished location, and their passion to preserve the memory of this site for future generations to keep the spirit of the Middlebrook Campground from being forgotten.

Author's Note: Grammar, spelling, and punctuation in direct quotes have all been retained from the original document, unless it was necessary to clarify meaning or intent with minor alterations.

Acknowledgments

Other than a few academic papers and archeological surveys, nothing significant or in-depth has ever been written about the Middlebrook Encampments. This work is the only known dedicated book that has been printed in the two-and-a-half centuries since its occupation by the American Army. Fortunately, two rich primary sources exist which have never been fully examined. All of the research used in this work has been derived from these sources—original military records, including the papers of officers at the encampments, and unpublished facts provided by local historians.

Several autobiographies, biographies, letters, and papers of American and British military leaders who were at Middlebrook provide astonishing details of events and many aspects of life at the encampments. Orderly books and muster rolls written during the time of the encampments provide a way to follow camp affairs almost day by day. The true image of this fascinating place is reflected in the eyewitness narratives found in these documents.

I was amazed to find several local historians with extensive knowledge of the subject matter, all of whom live near the encampments. These caretakers of local history provided special insights and information that cannot be found in recorded history. All of them specialize in a specific place encompassed by Middlebrook. They were eager to share treasured and often unknown details, and all showed a reverence for these cherished places and a passion to preserve these locations for future generations. They were willing to spend entire days trudging on foot with me through heavily wooded and precipitous terrain in the Watchung Hills. I especially value my adventures with these local historians: Arthur Sandor grew up on the site of the

Battle of Van Nest's Mill. The late Kels Swan devoted his life to the preservation of the tumultuous military history of Somerset County. Ray Sanderson, a reenactor and trustee of the Vanderveer House, is an expert on the events that occurred in Martinsville. The late Herbert M. Patullo, whose home stands atop the first Watchung Ridge on the site of Wayne's encampment, was a founder of the Washington Campground Association. Don McBride, whose home was adjacent to first ridge camp, rediscovered a rocky perch that likely served as a lookout. George W. Stillman, Sr. of the Edison Historic Association, has a profound knowledge of the obscure Battle of Short Hills. Jessie Havens, an accomplished journalist, contributed original information about the camp to local publications.

The staff and docents at all the superbly restored private houses that served as homes for senior officers during the encampments had stories to tell. Their extensive knowledge of the lives of private owners of the homes revealed many unfamiliar details of the social history, political affiliations, and culture of that critical time.

Fortunately, for historians studying the American Revolution, primary source information is available in major repositories. My research focused on the collections of the National Archives and Records Administration, the William I. Clements Library at the University of Michigan, and the David Library of the American Revolution, now housed in Philadelphia. The most rewarding research in original 18th century records came from scampering among the stacks of these comprehensive collections or viewing documents on microfilm.

Again, I extend my heartfelt thanks to "The Watchung Writers," an eclectic group of accomplished authors currently led by Pat Rydberg. These talented people, mostly fiction writers and poets, managed to stay the course while enduring my heavy doses of life in the Continental Army as they provided feedback on my early drafts. Their genuine interest and enthusiasm portends its appeal to future readers, since this book was written so that it can be enjoyed by the average reader not only serious military historians.

I thank my wife, Norma, for tolerating my obsession with the Revolutionary War and my virtual absences when I morphed into a time traveler and disappeared into the 18th century. Several members of my family accompanied me on my many visits and speaking events in the Middlebrook Encampment area. The spirit of Corporal John Allison, my muse and ancestor who served in the Continental Army for the entire eight years of that war, was with me during all of these adventures.

Table of Contents

Introduction i

Part I. Middlebrook and the War in Central New Jersey

1. Strategic Situation
   The War in Central New Jersey-January 1777..................3
2. Battle of Millstone
   A Militia Victory at Van Nest's Mill................................ 8
3. The Forage Wars
   Plundering the Jersey Countryside................................25
4. The Bloody Battle of Bound Brook
   Heroic Patriot Stand or Humiliating Defeat?...............35
5. The First Middlebrook Encampment
   May to July 1777 49
6. The Nest of the American Eagle..65
7. The Battle of Short Hills
   A Defeat that Saved the Continental Army...................78
8. The Rampaging Retreat
   The Redcoats Leave New Jersey 97

Part II. The Second Middlebrook Encampment, November 1778 - July 1779

9. Why Middlebrook? 107
10. The American Army Arrives
    The Winter Encampment of 1778-1779 From White Plains to Middlebrook 116
11. Settling In
    Hutting, the First Priority 127
12. Everyday Life
    For Soldiers at the Middlebrook Winter Camp...........134

13. The Women of Middlebrook..146
14. Festive Events and Fascinating Visitors
    Enliven Camp Routine 156
15. Crime and Punishment..167
16. Supporting a Deprived Army 176
17. General Greene's Worst Nightmare
    Supply Snags at the Camp..191
18. Civilian Support
    The Patriots of the Raritan and Washington Valleys 199
19. The Army Departs
    But the Tumult Continues, Middlebrook, 1779-1781................207

Part III. Exploring a dispersed Campsite in the Raritan and Washington Valleys

20. "Rank Hath Its Privileges"
    Officers Compete for Scarce Housing..220
21. Washington on the Rocks
    Green Brook, Middlebrook and Other Perches of the American Eagle 232
22. Scarce Research
    Ambitious Archeology, and a False Start 249

Part IV. The Artillery Camp in Pluckemin

23. With Knox at the Artillery Camp 262
24. A Well-Built Surprise
    An Academy and a Beacon..275
25. Our Compassionate General
    Headquarters at the Vanderveer House 286
26. Everyday Adventures at Pluckemin 293
27. Pluckemin Rediscovered Then Lost Again
    Archeology at the Camp... 301
28. Conclusion and Epilogue
    Middlebrook Remembered Today and the Future......................313

Appendices
A. Timeline of Middlebrook Campground and Related Military Activities 320
B. Deployment of Continental Army
   Winter, 1778-1779 323
C. American Generals Present at Second Middlebrook Encampment 324
D. Area Residents
   Who provided services during the encampment 325
E. Officer's Houses
   A Self-guided Tour 326
F. Sequence of Ownership Oliver Property 328
G. The Beacons on the Watchung Mountains 329
H. Somerset County Historic Sites 331
I. Somerset County Historic Associations 333

Endnotes 336
Bibliography 354
Index 364
About the Author 380

Introduction

THE REVOLUTIONARY WAR encampments of George Washington's Continental Army at Middlebrook and nearby Pluckemin, New Jersey, have been neglected by history. These places were critical to the American struggle during the Middle Atlantic campaigns. The highlands and surrounding valleys of Middlebrook, a natural fortress, were the location of two major encampments of Washington's Continental Army—during a harrowing seven weeks during the early summer of 1777, and throughout the course of the winter of 1778-1779. What is astonishing is that although the American Army spent close to a total of nine months in these locales, these hubs of the American Revolution languished in obscurity for over 200 years, during which time they virtually disappeared from national awareness.

    These encampments served as the centers of operations for American forces through much of the war, and during many of its darkest hours. Most significant is the site at Middlebrook, occupied during the winter of 1778-1779. Here the raw American Army matured into a cohesive fighting power capable of defeating the British forces, regarded at the time as the best trained and equipped army in the world. Unlike Valley Forge and Jockey Hollow, places that have been so eulogized that they are familiar to most school children, this sacred land, where decisive events occurred that changed the course of the war, is now built over by suburban creep. It is rarely marked, remains cloaked in mystery and mythology, and is slowly fading from memory.

    Middlebrook is located in present day Bridgewater Township. It includes the adjacent towns of Bound Brook and Martinsville,

New Jersey. The encampment derived its name from a stream that flows both through the Washington Valley and a gap in the first ridge in the Watchung Mountains at Chimney Rock. A small village of the same name was located at the junction of that creek and the Raritan River. This hamlet was eventually integrated into the town of Bound Brook and can no longer be found on modern New Jersey maps. Washington identified his correspondence with the name "Middlebrook" during the times he stayed there.

Beyond the encampments' basic function of sheltering soldiers, Middlebrook was a key component of Washington's Fabian strategy. Stationed on secure, mountainous terrain close to New York, the camp allowed the Continental commander-in-chief to monitor the enemy while avoiding direct engagement. This neutralized a numerically superior opponent while maximizing his own strength.

The growth in professionalism of Washington and his subordinates as military administrators while enduring the winter of 1778-1779 offers a telling new perspective on the commander's leadership during the Revolutionary War. It also demonstrates that winter encampments stand alongside more famous battlefields as sites where American independence was won.

Why did the Continental Army choose the Middlebrook site? Its centralized location meant that it could move in any direction to protect the state. It could threaten British-held New York and Staten Island, or quickly reenforce the vital Hudson Highlands. Its presence at this strategic spot dissuaded the British from crossing New Jersey by land to capture Philadelphia, the nation's capital at the time. This bastion in the mountains was the pivot in a defensive ring that the American Army created around the British headquarters in New York City.

The Watchung Mountain ridges run roughly in a northeast-southwest direction in northern New Jersey. This range stretches forty miles, from central New Jersey to the New York border. Merged into the Hudson Highlands in the north, the ridges formed a geological barrier and defendable band around New York City, the major British base in America. This high ground provided a 60-mile panoramic view from where the American forces could observe enemy troop movements and activities in the rich agricultural midlands below.

Observation posts and a beacon network on the heights, manned by Patriot defenders, could warn inhabitants across the entire state if the Crown forces invaded from the Hudson River or launched foraging raids into the interior. This impregnable natural fortification also provided a refuge if it became necessary to retreat into the swamps and forests of northwest New Jersey. The garrison at Middlebrook also encompassed the nearby colonial bridges stretching across the Raritan River in Bound Brook, all of which were vital to the Patriot cause.

During the spring of 1777, following a winter in Morristown, General Washington relocated his army of almost 10,000 soldiers into a new camp at Middlebrook in Somerset County. At that time the British army, consisting of 16,000 seasoned troops in the New Brunswick area, threatened to move west after easily routing the defenders of the outpost at Bound Brook on the Raritan River. The Americans first camped in the Washington Valley behind the first ridge, along the base of the Watchung Mountains on the fringes of Bound Brook. Here there was a fresh water supply and ample timber for construction and firewood. General George Washington stayed with Anthony Wayne's Brigade along the top of the ridge.

During this "First Middlebrook Encampment" General Washington developed a strategy of restraint by not knowingly engaging in a major battle when he knew that the odds were against him. This cautious approach led to an evasive action at the "Short Hills" in June 1777, in an area that is now encompassed by Edison, Scotch Plains and Plainfield. The British could not provoke a decisive engagement there, and the Patriots returned to the hills at Middlebrook. Incapacitated by a lack of supplies and forage, the Crown forces withdrew from the Garden State for the remainder of the war.

Introduction

The American Army returned to Middlebrook during the winter of 1778-1779. Its regiments spread out over what today covers the heights above Bound Brook, the Raritan Valley east of the ridges, and the Washington Valley behind the first ridge. The Artillery regiments set up a separate campground seven miles west at Pluckemin, in what is now Bedminster Township. During their winter at Middlebrook, Washington and his generals planned the massive Sullivan-Clinton Campaign against the Iroquois Nations.

Other units of the American army would be stationed separately at other strategic locations at the same time. The New Jersey Brigade was sent to nearby Elizabethtown to protect Raritan Bay and the New Jersey coast. Another brigade would be stationed at Danbury, Connecticut, from where they could move quickly to defend the Hudson Highlands or march south toward Manhattan if needed.

The first military training academy for officers was established in present-day Pluckemin, twenty-three years before West Point opened. This Middlebrook encampment may have been the site where the stars and stripes were first flown after a law to adopt a national flag was passed by Congress on June 14, 1777. The first training program for army surgeons was begun in Middlebrook and the Continental Army's light-infantry corps was also formed there under General Friedrich von Steuben.

Today much of the encampment terrain lies under highways and in heavily developed residential areas. However, a few extensive public tracts remain archeologically accessible. About 300 acres of this seasonal encampment have been acquired by the Somerset County Park System and the Washington Campground Association owns twenty-five acres in the heart of the encampment. Today, most of this protected land remains unchanged from the days of the Revolutionary War, but further development is likely. The grounds of the artillery campsite at Pluckemin are not currently accessible to the public, and there are no buildings or trails on this heavily wooded hillside. Any remaining land has been intensively built over with condominiums.

Archeological studies have revealed much about what occurred at sites in the main camp area at greater Bridgewater and in Pluckemin. Thousands of unearthed artifacts unearthed there both validate documented accounts and dispel misinformation concerning the activities at the camp and the composition of the Continental Army in 1778-1779. In the 1990s, the Somerset County Cultural and Heritage Commission retained an archaeological research firm to perform a study in the area occupied by Wayne's Brigade, near Middlebrook's Washington Rock. Ambitious work at Pluckemin, taking place over several years, has also uncovered a plethora of artifacts and structural remains.

Most people acquainted with Middlebrook are residents of nearby towns who faithfully attend a July 4th ceremony there each year. This event is held at a commemorative park which is maintained by the Washington Campground Association. In addition, the active "Friends of the Vanderveer House" sponsor several festive events at Pluckemin each year.

Until now, no single, readily obtainable source of information has been available about the encampments at Middlebrook. Records are mostly limited to statistically-oriented archeological studies and isolated military documents. The few academic papers that have been written are limited in scope and are not readily accessible. Other than these narrow sources, Middlebrook and Pluckemin are mentioned, if at all, in scant references buried in the broad context of standard histories of the American Revolution.

Unfortunately, the Middlebrook Encampment has not received the recognition it deserves and has never been regarded as a significant New Jersey tourist attraction. The objective of this work is to provide a broader understanding of these places, and to garner public attention by pulling many threads together to tell the story of the encampment.

In this book I examine the everyday lives of common soldiers, the social and domestic life of the officer corps, community relations, and the massive logistical burden that the army required.

Introduction

Uncovering fresh information from overlooked original documents and portraying the encampments in the larger context of the war in New Jersey revitalizes Middlebrook. My hope is that my efforts will trigger a deeper interest in this neglected and forgotten place.

PART I

Middlebrook and the War in Central New Jersey

Chapter One

Strategic Situation
The War in Central New Jersey
January 1777

THE ENCAMPMENT AT Middlebrook must be viewed in the context of the entire conflict that enveloped New Jersey during the Revolutionary War in order to appreciate its significance and the profound effect it had on the war's outcome. It was the hub of a prolonged fight for the control of the corridor from New York to the capitol of the new nation at Philadelphia. It also suppressed the struggle by the British to feed and otherwise supply their cumbersome army by pilfering the abundant resources of the state. For many months, the War for Independence was waged almost entirely in the Garden State and was driven as much by the need for provisions as by military strategy.

The British main army, numbering between 16,000 and 22,000 troops, and the American Continental forces, usually less than half that size, were always in close proximity to one another during this time. These opposing armies were commanded by the top military leaders in America. The Crown Forces were led by Sir William Howe and Lord Cornwallis. Commander-in-chief General George Washington led the Patriots. The highest ranking and most able commanders on both sides were present in New Jersey during this time. The American objective during the first six months of 1777 was to preserve the debilitated Continental Army by avoiding a major confrontation. A defeat at that time would surely have ended the war.

When the Continental Army left Trenton on January 2, 1777, the arrogant General Cornwallis declared, "We may easily bag the fox in the morning." He was wrong. The following morning Washington

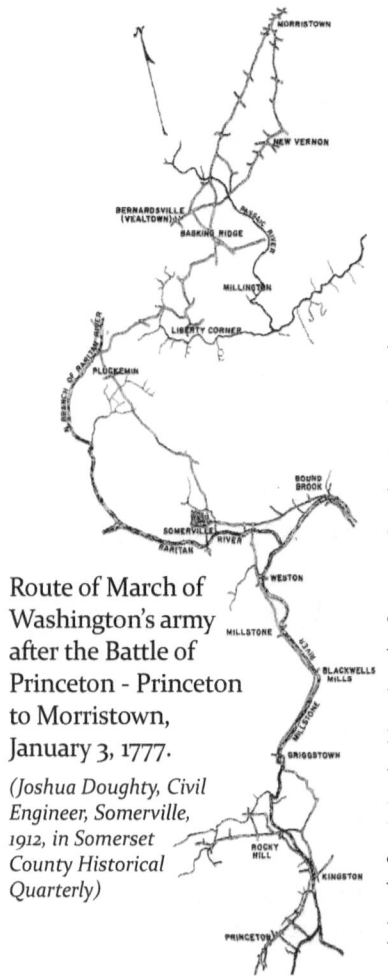

Route of March of Washington's army after the Battle of Princeton - Princeton to Morristown, January 3, 1777.
(Joshua Doughty, Civil Engineer, Somerville, 1912, in Somerset County Historical Quarterly)

attacked and defeated the British at Princeton.[1] In early January 1777, after the Patriot victories at Trenton and Princeton, both the British and Continental armies began planning for the traditional annual winter encampment.

After the victory at Princeton on January 3, 1777, the Continental Army in central New Jersey was at risk of being caught between the forces of Cornwallis in Trenton and the rest of the British Army at New Brunswick. Cornwallis and the remaining British and Hessian troops in and around Trenton and Princeton had withdrawn to New Brunswick after their defeat. Washington decided not to attack New Brunswick due to the exhausted state of his troops, Instead, he continued to move his army north up the Millstone River Valley toward Bound Brook and Morristown, New Jersey.

Strategically, Morristown was ideal for a winter encampment. It was about thirty miles from New York City, which would allow adequate time to prepare a defense if attacked or to reposition rapidly to counter British expeditions against the Hudson Highlands or Philadelphia. Geographically it was well protected by the Watchung Mountains to the south and the great swamp to the east. The Ramapo Hills, which joined the Hudson Highlands, offered protection to the north. The country was mountainous and densely wooded. Roads through this hinterland were rare and the few existing passes could be defended easily.

The Patriots headed north along the Millstone River and then marched through a gap in the high ground at Middlebrook. Today that pass is Chimney Rock Road in Bridgewater. They moved another seven miles to Pluckemin, where they rested for a few days before continuing twenty miles farther to Morristown, arriving there on January 6, 1777. Washington made his headquarters at the tavern of Jacob Arnold. This public house, which was on the Morristown Green, no longer stands.[2] The army encamped in the Lowantica Valley, now Woodland Avenue and Spring Valley Road.[3] The winter at Morristown was tranquil, but the camp was ravaged by a smallpox epidemic. It was always short of provisions, and food rations often dwindled down to only three days of reserve.

In January, while Washington's Army was secure in Morristown, the British and Hessian forces occupied a ten-mile corridor stretching from New Brunswick to Perth Amboy, including the neighboring villages of Bonhamtown and Piscatawaytown. The rich plains of central New Jersey extended between the opposing armies. This no man's land had abundant food, forage and supplies, all of which the Crown forces desperately needed to survive the winter. They immediately invaded these lush lands and began to ruthlessly pillage American farms. Their forays were stubbornly repulsed by American forces which largely consisted of local militia troops.

This period from January to June 1777 became known as the Forage Wars, since British raids were conducted to procure forage to feed their thousands of horses and obtain fresh provisions plundered from the bountiful countryside. American forces retaliated by conducting numerous scouting and harassing operations with units as large as 2,000 men. They quickly learned that skillfully planned guerrilla attacks were very effective against the strong but unwieldy Crown forces. Their combined losses in these skirmishes soon equaled those suffered in much larger battles. The Redcoat army was well positioned in the New Brunswick area, where they could either engage the Continental Army in a large decisive battle or withdraw back to their New York City headquarters.

The Crown forces gathered in New Jersey during this unusually tense time were on the defensive because their supplies were dwindling, and they were exposed to the ambushes of an aggressive New Jersey Militia. These committed American guerrillas were defending their own homes, farms and families in nearby villages. At first, British commanders were confident in the belief that the large supply of essential food and forage available from the rich New Jersey farmlands could easily sustain them. If they could hold six to twelve months of supplies in reserve, they could then resume an offensive campaign. But when their resources began dropping below the two-month level, British generals stopped planning offensive actions and instead began to contemplate evacuation.

Subsistence by foraging was never a successful tactic for the British. Many expeditions in the vicinity produced little or nothing of value. Civilians from whom supplies were taken were at first promised reimbursement but pillaging soon became the more typical method of procurement. Homes were burned and civilians killed. These raids alienated both local Patriots and the many Americans in the area who were either sympathetic to the British or were neutral.[4]

Although distracted by their attempts to survive the winter in New Brunswick, the British still maintained two strategic military objectives. First, they needed to lure the Continental Army onto open ground to engage it in a European style battle. Since they outmanned and outgunned the Patriots, the odds were in their favor that they could then win a final conclusive battle that would end the American rebellion. Second, the British Army wanted to cross the waist of the state over a secure land route to capture the Rebel capital of Philadelphia, the seat of the Continental Congress. They believed this would surely break the back of the revolt.

The intense fighting in New Jersey during the first six months of 1777 involved a series of skirmishes designed to stop the Redcoat foraging raids. It was punctuated by two battles—Bound Brook and Short Hills. Both were American defeats fought against overwhelming forces. But tenacious opposition by the citizen soldier of the New Jersey

Strategic Situation 7

Militia, supported by the regulars of the Continental Army, enabled the American cause to survive and continue the war. They effectively blocked the New York to Philadelphia land route, forcing the Crown forces to ultimately evacuate New Jersey. A contributing factor is the fact that the British were reluctant to penetrate the daunting mountain ridges, swamps, and forests of western New Jersey.

The focal point of all this action was Middlebrook, a natural fortress on the high ground of the first ridge of the Watchung Mountains. In 1777 the campground at Middlebrook was a secure location. As mentioned, it allowed General Washington to conceal his army and to exercise his strategy of restraint by not engaging in a major battle when the odds were against him. This cautious approach, combined with their unsuccessful foraging efforts, finally drove the British forces to leave the state. It also prolonged the war by debilitating the enemy. The Middlebrook Encampment later served as the winter encampment for the American Army during the winter of 1778-1779, and as the operations center for the protection and reoccupation of the state for a significant part of the war.

Braving the frigid winters at camps is the iconic image of the American Revolution. These camps were a critical factor in the waging and winning of the War of Independence. Exploring the inner workings of the Continental Army show how camp construction and administration played a crucial role in Patriot strategy. During the war, troops generally spent only a few days a year in actual combat. The rest of the time, especially in the winter months, they were engaged in a personal battle against a combination of elements, unfriendly terrain, disease, and hunger. Victory in the sustained struggle depended on a mastery of camp construction, logistics, and health and hygiene. Environmental, administrative, and operational investigation of the winter encampment at Middlebrook is a worthy exploration.

Chapter Two

Battle of Millstone

A Militia Victory at Van Nest's Mill

The First Incursions

ON JANUARY 20, 1777, a large British foraging party of about 500 men left New Brunswick and headed west toward the Millstone River. They eventually reached Van Nest's Mill, a major American supply depot located at Weston, New Jersey. Today, the mill site is on the eastern bank of the Millstone River across from present day Manville, where the Wilbur Smith Bridge crosses the river. A battery of Hessian cannons was positioned nearby to protect the raiders. While at the mill these foragers seized a large number of cattle, sheep, and horses, as well as several wagonloads of flour and other supplies.

As the British wagon train prepared to return to New Brunswick the Patriot forces of General Philemon Dickinson waded through the icy waist-deep water of the Millstone River and attacked them from their rear. The American musket fire halted the wagon train and silenced the Hessian cannons. After taking heavy losses, the surprised Redcoats fled the area empty-handed. This fierce clash, neglected in history, has been called the Battle of Millstone or Van Nest's Mill.

The Forage War had begun on January 3 when an advanced British supply column began marching north from Princeton, only a day after the Crown Forces were defeated there, toward New Brunswick. Near Ten Mile Run, a tributary of the Millstone River, the British stumbled into onto the Somerset Horse, a New Jersey militia cavalry unit commanded by Captain John Stryker. The Americans surrounded and captured the British wagons and promptly sent five of them loaded with warm woolen clothing to Washington's freezing troops huddled around campfires in Morristown. Captain Stryker was killed later that year at the Battle of Brandywine.

General Washington was encouraged by signs of British apprehensiveness. He considered the frantic British foraging to sustain their base at New Brunswick to be major evidence of their distress. Their plans to defeat the Continental Army in a final decisive battle and occupy New Jersey, a critical crossroads state, could not proceed without provisions and forage.

On January 17 Washington wrote that the British were "endeavouring to draw in all the Forage they can get, in the course of which, they have daily Skirmishes with our advanced parties." He wrote to Major General Joseph Spencer that "the Enemy, by being drove back from most part of the province of Jersey, on which they depended for Subsistence, are much distressed for Provision and Forage, and unless they make a push to extricate themselves, they must in a Manner perish this Winter."[1]

Charles Stedman, a British commissary officer, verified the desperation of the British army camped around New Brunswick. In his account he commented that while the Crown forces indeed did have vastly superior numbers, they were confined to a narrow belt of land between New Brunswick and Perth Amboy. He confirmed that whenever they ventured out in their quest for provisions, they suffered great losses in skirmishes with the Americans based on a line of outposts that extended from Morristown to Woodbridge. The goods that the Redcoats seized were often recaptured by parties of the raiders, who were alerted by the numerous Patriot civilians who lived in the area.[2]

On January 13, a detachment consisting of several hundred British and Hessian scavengers advanced from New Brunswick and headed eight miles west to Somerset Court House, today's Millstone, New Jersey. After staying in the village for about a week they withdrew back to New Brunswick. Along the route they burned houses, plundered property, and stole livestock.

This bold incursion caused New Jersey militia companies to join Colonel John Scott's Continental troops in setting up major outposts at Millstone and Bound Brook. From that time onward, the area west of New Brunswick up to the Millstone and Raritan Rivers was considered by both sides to be a no man's land.

The Dismal Swamp north of Metuchen flows west to join Green Brook, and the combined streams empty into the Raritan River at Bound Brook. Along the length of these watercourses the only bridges were at Samptown (now South Plainfield) and its neighboring village of Quibbletown (now New Market), one mile to the west. The American front line of outposts was established along these natural defenses.

British patrols scouted the area during the first two weeks of January 1777 without meeting resistance from the Americans. Hessian Captain Johann reported that the freezing weather also hindered the arrival of provisions: "The 33d and 42nd regiments with a battalion of the 71st Regiment and some companies of light infantry were stationed at Bonumtown and Piscataway to keep that communication open, for the river was frozen up, so that the provisions could not be brought from Amboy to Brunswick by water, the most part of the winter."[3]

At that time, the Patriot forces in the area consisted of 600 troops from the 1st Connecticut Militia and the 8th Virginia Continental Line. They immediately dispatched patrols to molest the British outposts along the Raritan River. At this point both sides were in dire need of supplies. Colonel Scott, the American commander, reluctantly gave orders to strip all the farms between the Watchung Mountains and the Raritan River for desperately needed food, forage, livestock, and wagons. In the days ahead foragers from both armies roved through the area, often exchanging gunfire. The Americans were successful in gathering the most supplies first. Lacking the transportation to remove larger quantities, they began to stockpile them at depots at Millstone and Samptown.

Washington wrote to the Continental Congress on January

12, 1777, "...the Enemy have made no Move since my last, by every Account they begin to be distressed, particularly for Forage, of which there is little or none remaining in the small Circle they possess." Two days later he again reported from his headquarters in Morristown that the British were at "Amboy and Brunswick," and "our Accounts still confirm their want of Forage, which I hope will increase."

The Battle at Van Nest's Mill

An extraordinary amount of Revolutionary War activity occurred on the road between North Jersey and Princeton, and between Princeton and New Brunswick. This was in the center of an area frequented by British troops foraging from New Brunswick. As a result, the citizen soldiers stationed near the Somerset Court House were constantly in action.

As mentioned, on January 20, sixteen days after Washington had passed by Middlebrook with his victorious army after the Battle of Princeton, a large British foraging party of about 500 men, led by Lieutenant Colonel Robert Abercromby of the 37th Foot, left New Brunswick and headed west toward the Millstone River. They crossed over the Raritan River, but it is unclear exactly which bridge they used. A rear guard of Hessians with a battery of field artillery was left behind to cover the bridge. They eventually reached Van Nest's Mill, the major American supply depot, a few miles north of the Somerset Court House near the point where the Millstone River empties into the Raritan. At the mill they seized cattle, sheep, 100 horses, and 49 wagonloads of flour and other supplies. They then prepared to return to New Brunswick.

Captain Thomas Rodney of the Delaware Line, who commanded the van of the army on the march toward Somerset Court House:

We then marched on to a little village called Stone Brook or Summerset Court House about 15 miles from Princeton where we arrived just at dusk. About an hour before we arrived here 150 of the enemy from Princeton and 50 which were stationed in this town went off with 20 wagons laden with clothing and Linen, and 400 of the Jersey Militia were afraid to fire on them and let them go off unmolested, and there were no troops in our army fresh enough to pursue them, or the whole might have been taken in a few hours.[4]

American militia companies in the north were alerted to this British incursion early in the day, and quickly marched toward Bound Brook, Millstone, and Van Nest's Mill, About 400 New Jersey and 50 Pennsylvania soldiers formed under Brigadier General Philemon Dickinson to repel the British action. Historian Arthur Shandor notes that the Patriot militia was stationed in the area of the Weston section of Manville in the area of Onka Street, Whelan Street, Frech Avenue, and Railroad Avenue in South Manville. They were also positioned on the South Side of Manville in the area of Roosevelt Avenue, Camplain Road, (Camp Lane), and South 9th Avenue to South 11th Avenue. This is an area that reaches past today's Manville Public Library to the present Lehigh Valley Railroad tracks and into the North Side of Manville.

While detailed accounts of their movements are sketchy, Dickinson apparently divided his forces, sending one force to meet the front of the British wagon train, while a second moved to flank them. This last force forded the Millstone River, wading in icy water that was waist deep. They surprised the British wagon train in the lane near the mill before it reached the main road and the bridge to return to New Brunswick.

American musket shots immediately struck horses hauling the first wagon. This stopped the train, scattered the wagon drivers, and drove the British into retreat, leaving their booty behind. When the militiamen reached the bridge two miles north at Millstone, the Hessian rear guard fired grape shot from its artillery to cover their flight. After an exchange of fire across the river without any apparent consequence, the British fled.

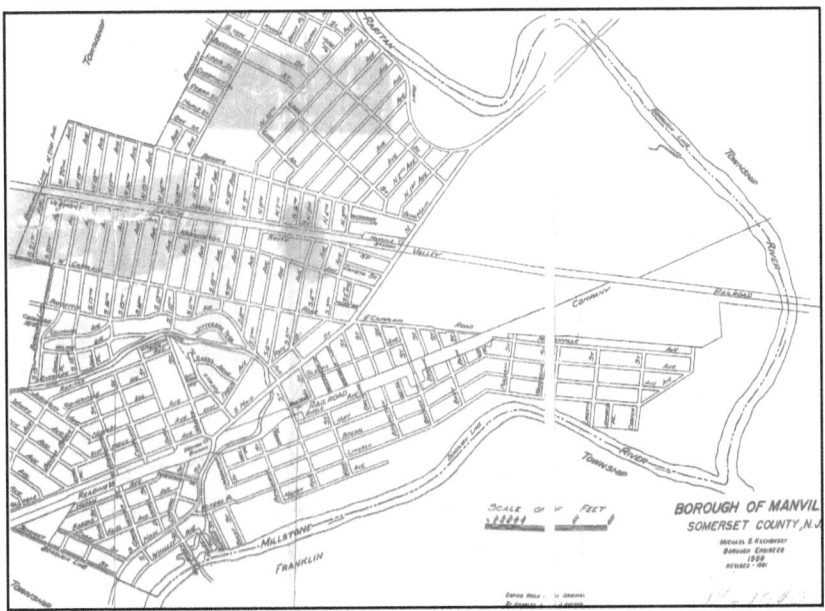

Borough of Manville-1958. *(Michael Kachorsky, Borough Engineer, 1958. Courtesy of Arthur Shandor, Manville, New Jersey)*

The triumphant Dickinson wrote in a letter to Colonel John Nielson on January 23, "I have the pleasure to inform you that on Monday last with about 450 men chiefly our militia I attacked a foraging party near V. Nest Mills consisting of 500 men with 2 field pieces, which we routed after an engagement of 20 minutes and brought off 107 horses, 49 wagons, 115 cattle, 70 sheep, 40 barrels of flour—106 bags and many other things, 49 prisoners."

General Washington, who was not always pleased with the performance of the local militia, to John Hancock on 22 January 1777:

> *My last to you was on the 20th instant. Since that, I have the pleasure to inform you, that General Dickinson, with about 400 Militia, has defeated a foraging Party of the Enemy of an equal number, and has taken forty Waggons and upwards of hundred Horses, most of them of the English draft Breed, and a number of Sheep and Cattle which they had collected. The Enemy retreated with so much precipitation, that General Dickinson had only an*

opportunity of making nine prisoners, they were observed to carry off a good many dead and wounded in light Waggons. This Action happened near Somerset Court House on Millstone River. Genl Dickinsons behaviour reflects the highest honour upon him, for tho' his Troops were all raw, he lead them thro' the River, middle deep, and gave the Enemy so severe a charge, that, altho' supported by three field pieces, they gave way and left their Convoy, and only reported the taking of nine prisoners.[5]

British officer Archibald Robertson's diary entry for January 20, 1777 agrees with Washington's account of the skirmish on that date: "Lieutenant Colonel Abercromby with 500 men went on a foraging party towards Hillsborough. Part of this Corps was attacked by the Rebels, which occasion'd such disorder Amongst the Waggon Drivers that 42 Waggons were left behind."[6]

Historian Benson Lossing provided an interesting summary of this action in his *Pictorial Field Book of the Revolution*:

A line of forts had been established along the Millstone river, in the direction of Princeton. One of these, at Somerset Court House, (the village of Millstone), was occupied by Gen. Dickenson with two companies of the regular army, and about 300 militia. A mill on the opposite part of the stream contained considerable flour. Cornwallis, then lying at New Brunswick, dispatched a foraging party to capture it. The party consisted of about 400 men and more than 40 wagons. The British arrived at the mill at Weston, in the morning and having loaded their wagons with flour, were about to return, when Gen. Dickenson leading a portion of his force through the river, middle deep, and filled with ice, attacked them with so much spirit, that they fled in haste, leaving, the whole of their plunder with their wagons, behind them. Dickenson lost five men in this skirting, and the enemy about 30. Washington warmly commended Gen. Dickenson for his enterprise and gallantry evinced in this little skirmish.[7]

Eyewitness Accounts

Samuel Sutphen, a Slave

Samuel Sutphen, a local slave, lived near the Millstone riverbank and was familiar with the local terrain. According to details in his government pension application, he led Dickinson's American defenders to the place where they could wade across the Millstone River to attack the raiders, a move that may have led to the Patriot victory. Sutphen was sent by his master to the local militia company as his substitute. It was a common practice for slaves and indentured people to serve in place of their owners. Typically, they were promised freedom when the war ended, but these types of agreements were usually broken.

Sutphen gave his lucid account of the attack on Van Nest's Mill to a court clerk when he applied for a pension in 1834:

> A party of Hessians, about 1 company (70), an escort for these teams from Brunswick, was discovered secreted behind a hedge with some 4 or 5 field pieces. They fired upon us and retreated. We followed on a piece, but Lt. Davis ordr'd us to retreat. Davis' Capt. Westcott from Cumberland had been left sick at Guysbert Bogert's, where he died, and was taken back to Cumberland Co[unty]. There was a large body of militia out, and Gen'l Dickinson commanded. The firing was principally across the river at the bridge. I was out on this alarm but one day. We mounted guard along the branch above the 2-bridges almost every night; nearly all this winter and spring on guard duty.[8]

His testimony was given fifty-seven years after the action, when he was eighty-seven years old. His application confirms that he was a slave during the duration of the war. His request for the pension was denied for lack of detail, in particular his time of service and absence of documentation. In contrast, white veterans made simpler, less detailed applications but were approved. As a slave, he was not able to volunteer or enlist. Only those who served voluntarily were eligible to receive a pension, so in a sense his application attempt was a moot point. Samuel was illiterate and did not know which name he had been enrolled under; this application shows he was called Sutphin only after the war. He cannot be positively identified on the Continental records by the War Department. These documents list several men identified only by the given name Samuel.

William Churchhill Houston - Militia Officer

Captain William Churchill Houston was a mathematics professor at Princeton University who was commissioned by the Somerset County Militia. A number of his students joined his regiment. Author Thomas Glenn suggests that the following excerpt from a journal may have been written by an anonymous soldier serving under Houston, possibly one of his students. It describes the failed attempts of the Americans after the action at the mill, when they attempted to pursue the British raiders as they fled back to New Brunswick:

Staid here in peace till Monday morning [January 20] we then received an Alarm and were ordered to march to Boundbrook, we arrived there between 11 and 12, then hearing that the Enemy was plundering at Millstone, we immediately marched for that place, being joined by a considerable body at Boundbrook we marched on till we passed the bridge [Van Veghten Bridge, Finderne Avenue] hearing several Cannon fired, while on the way. After crossing the Bridge, the Battallion I was in was taken off for the left wing, I crossed Millstone, some distance below the Bridge, wading through the water, more than knee deep. We immediately marched towards the road, and fired upon the Baggage Guard, who were retreated that way. They immediately left horses, wagons and plunder, and returned with the greatest precipitation. The main body of the Enemy lay just over south of the bridge. Before we crossed the River below, our main Body began the Attack at the Bridge with one Field piece and made the Enemy give way. They continued their fire upon the Enemy some time. Our wing, after driving the Baggage Guard, pursued on and flanked the Enemy. After a short engagement, finding ourselves greatly overpowered with numbers, we received General Orders to retreat, having had 1 man killed and 2 wounded. and we had taken 2 of the Enemy prisoners. We then retreated back to the River, lest our retreat should be cut off. But finding the Enemy did not pursue, we rallied again, with as many of our men as we could collect, and marched on towards the Enemy the second time; but when we came in sight of them, they got possession of an eminence in the End of a clear Field, with one or more Field pieces and poured down their Grapeshot upon us briskly. Then finding it in vain to attack them with our little Body, under so great a disadvantage, we immediately retreated back and most of our men went over the River up into a clear field, to where our main Body had by this time collected....[9]

Thomas Sullivan - British Private

Thomas Sullivan, a private in the British foraging party, also provided an account of the engagement at Van Nest's Mill. He grossly inflated the size of the American force, and almost casts the event as a Redcoat victory.

January 21 A detachment of 100 British Grenadiers, 100 light infantry, 200 Hessian Grenadiers and a squadron of Light Dragoons with 2 three pounders under the command of Lieut. Colonel Abercrombie went from Brunswick with all the wagons of the army, about 9 miles from town toward the Bridge that was on the Raritan [Millstone] above Hillsborough. Major Dilkes with 100 British Grenadiers marched in the rear of the foragers, and took post in the skirt of the woods on their left, having the river on the right.

There were about 4,000 of the Rebels that mustered out of the woods, that attacked ye front of the wagon line and drove off 24 of the English wagons with four horses in each before the Grenadiers could come up. Major Dilkes with his party engaged them with two field pieces and kept his fire up, until they expended all their ammunition, at the rate of 60 rounds per man. Then they retreated to the second party of Grenadiers from whom they got more ammunition.

During the interval the Hessian Grenadiers with their two pieces of the cannon attacked ye enemy upon their flank and kept them in continual Play, until the British Grenadiers and Light Infantry joined them. The engagement began very hot but with their united force and unusual bravery they repulsed the enemy, driving them across the bridge which they defended for some time. The Foragers threw their forage away mostly and made the Best of their way home. Our loss in the action did not exceed 12 men killed or wounded.[10]

The Millstone is a small river, usually averaging a couple of feet in depth and fifty feet in width. It begins in Monmouth County, meanders through Somerset County, and eventually flows into the Raritan River at Manville, New Jersey. A crossing on foot in the icy water in the middle of the winter, when the water was higher, followed by a sharp skirmish, is a testimony to the determination of the American troops. The aggressive behavior shown by the militia demonstrates that a change had occurred after the battles of Trenton and Princeton. The belief in the invincibility of the British regulars had been shown to be false, and the inhabitants of New Jersey were not going to be robbed of their sustenance without stiff resistance. British resources strained to increase the size of units protecting themselves during foraging trips.

Battle of Millstone

Local Historian Arthur Shandor Locates Sites of the Battle of Van Nest's Mill

Arthur Shandor grew up in Weston section of Manville, New Jersey. His home for many years on Fable Avenue was a few hundred feet from the scene of the engagement at Van Nest's Mill. He related this unusually thorough narrative to me about his exploration of the area. I believe it is the only contemporary account of that historic event that exists. He also provided photos and maps and gave me an informative tour of the battle area. Mr. Shandor, a retired educator, resides in Hillsborough, New Jersey and offers these comments:

Local Historian Arthur Shandor on the Millstone River, with Wilbur Smith Bridge and Van Nest Mill site in background.
(Author's collection)

I want to offer up thoughts to you about the engagement at Van Nest's Mill in the Weston section of Franklin Township on the east side of the Millstone River. I went to Google maps and did an aerial view of the Millstone River from near Sacred Heart Cemetery, going north along the river to where the Millstone flows north into the Raritan to determine the location of the crossing of the American force of the Millstone River north of the Van Nest's Mill.

I'm very familiar with the terrain from the Sacred Heart area all the way to the point where the Raritan meets the Millstone. If you do an aerial view of the Millstone and view the farmland on the eastern side of the Millstone you can see how flat it is. If you scan it from Wilbur Smith Bridge all the way to the Raritan, you can see how the terrain along the shore changes. Starting with the mouth of Royce Brook where it empties into the Millstone and then continuing north you see a high parcel of land that is flat. That used to be Selodies sod farm when I was a kid. Continuing north along that flat parcel of land there is a municipal park. Where the flatness ends you pick up Lincoln Avenue, which runs east towards the river. The property that slopes down to the river where Bank Street meets Lincoln Boulevard has a steep slope and a big drop.

From that point on, going north towards where the Millstone River meets the Raritan River, the riverbank has steep drops. As we can surmise, the troops would not want to wade across the Millstone River where the western bank drops steeply. I believe they crossed at the northern edge of the municipal park where there is no steep slope to the river which is at the northern end of the Lincoln Avenue Children's Park and about 1,500 yards from the Hessian cannon emplacement that defended the bridge and mill.

The Hessians likely had smaller, more mobile cannons similar to the American three-pounder "galloper." The effective range of a three-pounder was 750 to 1,000 yards. Therefore, the river could be safely crossed at that point without harassment from artillery if they were detected. The river here is about 25 yards wide with low banks on either side. The land between the railroad and the Millstone River in this area is called the Lost Valley section of Manville.

Using map scale, I measured the straight-line distance from the cannon's site in Little Weston to Bridge Street as it intersects Huff Avenue in the Lost Valley. The distance is about 4,400 feet, roughly 8/10 of a mile. The length of the children's park on Lincoln Ave. (and Selodie's sod farm) is roughly the same length. Royce Brook, of course, runs between the site of the hill and the southern side of the farm and park. Knowing the estimated effective range of the three cannon, one could reasonably guess how far away the militia would have to be in order to cross the Millstone River without being on the receiving end of the British cannonade. The

British set up the three cannon on a nearby hill to fire upon the attacking militia if they attempted to attack the foragers looting the mill. I know where this hill used to be. Locate South Main Street and Van Nest Place, Central Street, and Fable Ave. In this triangle there once was a huge parcel of land that jutted out towards Wilhouske Street. This was once the Rudovick property which sat on a hill. I believe this is where the British set up their cannon.

There were no other hills in the vicinity. In the late 1960s, this hill was excavated out so that the curvy South Main Street could be straightened out to accommodate four lanes. The Abraham Van Nest property encompassed this small hill which reached out to the great road, now a portion of South Main Street. Old maps depicted two little villages, one on the river by the mill and a few buildings on the river by Central Street and Fable Avenue. Much evidence of military activity must still lie buried in this area.

What is now Rt. 533, paralleling the Millstone River from Griggstown area up to and beyond the Van Veghten house area was known as the Great Road. The location of the original Van Nest's mill has been debated. Was it originally at the site of the present day ruins by the Wilbur Smith Bridge or where it appeared on a 1776 map, 1/2 mile south of the Wilbur Smith Bridge, on the east bank of the Millstone River, across from the entrance to the present Central Jersey Regional Airport.

Hendrik Schenk, a Dutchman, built the original mill about 1740. We know for sure, the timbers of Schenk's original mill, were used to build the newer mill by the Wilbur Smith Bridge in 1844. There is no evidence that the timbers were transported to the present site from another location. Surely there would be records of who was hired to dismantle the first mill, who hired the wagons and the horses to cart the materials to the present site, receipts of how much it cost receipts for payment of bills, etc. No relics of the battle or any accoutrements were ever found at the site, nor has a sluice been found there.

Author's note: A map of the British outposts between Burlington and New Bridge, New Jersey in December 1776 shows a place identified as "Vaness" near the junction of the "Millstone Creek" and "Bound Brook," as well as what is the Raritan River. A bit further to the west, along the Millstone Creek, is also identified as "Schenk."[11] Johann Ewald's map, "Plan of the Area of Bound Brook, 20 April 1777," does not identify the Van Nest's Mill by name, but does show a mill at the location of what seems to me to be the Wilbur Smith Bridge, before the intersection of the two rivers.[12] The river banks north of the bridge have also been thoroughly explored on foot by local residents without finding any evidence of a building site. While no artifact evidence has been found at the mill site at the bridge, aside from the 1776 map which seems to be approximate, all other evidence indicates that the mill was always by the Wilbur Smith Bridge, where its remnants still exist. Shandor continues:

> *Abraham Van Nest, the owner of the mill, had a farm with a homestead and outbuildings that occupied the land directly across the river from his mill. The farm was between what is now Van Nest Place, Central Avenue and Fable Avenue. This area encompassed the small hill where the Hessian Cannons were placed.*
>
> *The Patriots militia was stationed in on the South Side in Manville around the intersection of Railroad Avenue and Onka Street and in Weston a little less than a mile from where they forded the Millstone River. What route did they take to reach the river crossing site? As we know the present physiography in Lost Valley Section and Weston Section in Manville can thwart an amateur history sleuth in regard to the battle at Van Nests Mill. When the Philadelphia and Reading railroad was built in the 1870s, an elevated embankment was built to accommodate the railroad tracks to remain on the same plain.*
>
> *Of course, this embankment and railroad were not there in 1777, so the troops from the south side could quick march directly eastward and cross the Millstone to to the flat farmlands*

to the east side of the Millstone River and come up behind the British at the mill. The American militia on Onka St, Frech Avenue, and Whelan St., joined up with the others who had been encamped on the south side in Weston and then crossed the Millstone.

A map of the British outposts between Burlington and New Bridge, New Jersey in December 1776 shows a place identified as "Vaness" near the junction of the "Millstone Creek" and "Bound Brook," as well as what is the Raritan River. A bit further to the west, along the Millstone Creek, is also identified as "Schenk..[13] Johann Ewald's map, "Plan of the Area of Bound Brook, 20 April 1777," does not identify the Van Nest's Mill by name, but does show a mill at the location of what seems to me to be the Wilbur Smith Bridge, before the intersection of the two rivers.[14] The river banks north of the bridge have also been thoroughly explored on foot by local residents without finding any evidence of a building site. While no artifact evidence has been found at the mill site at the bridge, aside from the 1776 map which seems to be approximate, all other evidence indicates that the mill was always by the Wilbur Smith Bridge where it remnants still exist.

Interesting Revolutionary War Era Places in the Millstone Valley

Other nearby historic sites include the American spy John Honeyman's house, located at 1008 Canal Road in Griggstown, and the Van Doren House in Millstone, where Washington lodged after the Battle of Princeton. You can also visit the site of the Somerset Court House in Millstone which was burned down by Simcoe's raiders, and the site of the Cornwallis encampment in Colonial Park, near East Millstone.

A 300-year-old blacksmith shop and a Dutch home are located by the river in Millstone behind a present day tavern/liquor store. The Old Millstone Forge is home to what was the longest operating blacksmith shop in America. It served the area as an active blacksmith's shop until the death, in 1959, of Edward H. Wyckoff, its last blacksmith, who served a remarkable sixty-four years. The building was restored by area residents in the 1960s and was until recently was again operated as a blacksmith shop and living history museum by the Millstone Forge Association. On the corner of Old Amwell Road and Route 533 stands the Hillsborough Dutch Reformed Church, built in 1766. Abraham Van Nest, the second owner of Van Nest's Mill and the owner on January 20, 1777, is buried in the adjacent churchyard with his wife Catherine.

The Battle of Millstone or Van Nest's Mill was the first of more than 100 skirmishes in the spring of 1777. The incessant combat beneath the Watchung Mountain from New Brunswick, Perth Amboy, Bound Brook and north along the Short Hills tested the strategy and endurance of both the Crown forces and the American army.

Ruins of Van Nest's Mill. *(Author's Collection)*

Chapter Three

The Forage Wars

Plundering the New Jersey Countryside

THE EVENTS THAT lead to the first encampment at Middlebrook in May 1777 began with the iconic crossing of the icy Delaware River by General George Washington's army on the night of December 25-26, 1776. Over the course of the next ten days the American forces won two critical battles. Washington surprised and defeated a daunting outpost of Hessian mercenaries in Trenton on December 26, and a week later a daring night march led to the victorious Battle of Princeton. These defeats of an enemy which the Americans had considered invincible preserved the morale and unity of the Continental Army. American control of much of New Jersey was reestablished, and the waning spirit of the entire nation was well on the road to recovery.

Curiously, after the American triumphs at Trenton and Princeton, neither side followed up with a counterattack. Instead, they both choose the 18th century tradition of seeking a secure place to settle for their winter encampment. Flushed with the successes of Trenton and Princeton, George Washington wanted to push onward to New Brunswick to raid the British munitions stores there and attempt to drive the British army out of New Jersey entirely. But his officers were more cautious and realistic. They convinced him to move the exhausted Continentals to a safe winter bivouac. Morristown, a secure locale situated behind mountain ridges and a great swamp, was selected as the best site for the base. Washington later wrote, "The Severity of the Season has made our Troops, especially the Militia, extremely impatient, and has reduced the number very considerably. Every day more or less leave us. Their complaints and the great fatigues they had undergone, induced me to come to this place, as the best calculated of any in this quarter, to accommodate and refresh them."[1]

General Howe's Strategic Errors

The British were completely unaware of the extreme weakness of the American army at the time. It has been reduced by rampant desertions, the unreliable nature of the local militia forces, and a severe smallpox epidemic.[2] William Howe missed an unusual opportunity by choosing not to pursue Washington into western New Jersey, despite his vastly superior army. He had little doubt that he would vanquish the Patriots by the spring. His overconfidence led Howe to commit strategic errors. He did not anticipate the intensity of insurgent guerrilla-type warfare that soon occurred, and he did not expect the Americans to stubbornly avoid a large definitive battle.

By the time Howe was finally ready to resume his New Jersey campaign in June 1777, the tide had turned. General Washington had been supplied with French weapons, and the Continental Army had been reinforced by strong enlistments and the addition of hard-hitting militia brigades. The Patriots now had the ability to harass the British foragers at will without being drawn into a major battle.

Foraging Becomes Thorny for the British Occupation Forces

During the first half of 1777, after the Battle of Millstone at Van Nest's Mill, the violent skirmishing north and west of New Brunswick in the Raritan Valley and along the Millstone River continued. The intense conflict caused by British raids for supplies to maintain their vast army of 16,000 troops, civilian support personnel, and the local Loyalist population was a daunting challenge. The Americans continuously assaulted enemy foraging parties and the British front lines. The action centered on Quibbletown and Samptown. Quibbletown today is Piscataway Township. Samptown, one mile west, is today's town of South Plainfield.

This was the Raritan River Valley. It extended north and west of New Brunswick to the Millstone River Valley. It was a bountiful flat land with rich farms ripe for plundering. The massive effort to gain supplies and limit the enemy's access to provisions was the focus of all military operations in New Jersey during the first half of 1777. This relentless enemy activity finally led Washington to move the Continental Army from Morristown in May 1777 to Middlebrook, to be near to the front lines. Called the First Middlebrook Encampment, this lasted for the next seven weeks.

The Local Militia Awakens

The Washington Valley lies between the first and second ridges of the Watchung Mountains in Somerset County. It encompasses the present-day towns of Watchung, Warren Township, Millington, Long Hill, and Martinsville. These towns were passionately patriotic in 1777, and throughout the duration of the war. Nearly the entire male population of New Jersey served either in General Washington's Continental Army or as minutemen in the state militia during the eight years of the Revolutionary War. Militia men in central New Jersey were stationed at the Blue Hills post on Vermeule's Farm in North Plainfield, guarding the mountain passes at the Quibbletown Gap in Greenbrook and Stony Brook pass at Watchung. They also fought bravely at the major engagements of Springfield, Connecticut Farms, Short Hills, Bound Brook, Monmouth Court House, Long Island and White Plains, as well as at scores of smaller skirmishes. Many served as infantrymen, while others were fifers. drummers, teamsters, express riders, and scouts or spies.

Local men who enlisted in the Continental Army were typically signed up by recruiters at the town's tavern. Recruiters typically were handsome, immaculately-uniformed young officers selected for their appearance and charisma. They often carried a flag and were accompanied by a fifer and drummer in order to attract attention. They gave stirring patriotic speeches explaining that it was every man's obligation to protect their homes and loved ones. The potential recruits, many still naive teenagers, would then be invited into the village tavern where an unlimited supply of free liquor would be provided. Wide-eyed farm laborers were easy pickings for the recruiters. They signed up as much for adventure and the promise of a uniform and pay, as for patriotism.

The abundant farmland in this area stretched from the Raritan River twenty miles northeast to Springfield. The entire expanse was at the mercy of Redcoat foraging parties from New Brunswick who dominated the eight miles across the Millstone River as far west as Neshanic and the South Branch. The ravenous raiders stole everything they could lay their hands on, and murdered any farmers who offered resistance. These marauders often suffered heavy losses inflicted by the enraged citizen soldiers of local militias.

Fighting Accelerates at Quibbletown and Samptown

The New Jersey Militia was encouraged by its early success at Van Nest's Mill, so it quickly followed up three days later by boldly attacking two British regiments near New Brunswick on January 23. General Cornwallis was infuriated by an effective strike so close to his base by these part-time soldiers, and he attempted to retaliate by sending 600 of his best troops after them the next day. This detachment consisted of light infantry, cavalry, and a Hessian Jaeger battalion. Their orders were to seize American supplies stored at Samptown.

This taskforce marched the ten miles from Raritan Landing in New Brunswick to Piscataway. On the way they looted farms along present day New Brunswick Avenue and New Market Road in South

Plainfield.³ The New Jersey Militia attempted to repel the invaders, but were unable to stop the large column of elite professional soldiers. The outnumbered American force was driven off near Quibbletown, but the Redcoat success cost them thirty to forty dead and several wounded.

In his journal, British Private Thomas Sullivan boasted about the performance of his comrades at Quibbletown:

> *January 23d - The 29th and 35th Battalions with an attachment of Royal Highlanders under the command of Lieutenant Colonel Prescott went to cover the provision wagons, being halfway between Brunswick and Amboy where a large party of rebels advanced from the woods upon them with three pieces of cannon. The Highlanders being drawn up on an eminence, attacked the enemy on the edge of a wood and advanced to them under heavy fire from the enemy. The Highlanders observing that the Rebels would not advance out of the wood, made a charge upon them, which was always a terror to the rebels, and put them to immediate rout. The enemy could never endure to stand for any time to the Bayonet. But if the King's troops kept at a distance, they stood firing with musketry long enough.*⁴

Three days later, on January 27, another unit of 600 British infantry, 50 calvary, and a Hessian grenadier battalionmarched out again from Raritan Landing. The Hessians from the Kessel Company of Jaegers followed their orders to seize supplies stored in the vicinity of Samptown. Under command of British General Leslie and Hessian Colonel Von Donop, these troops looted farms along present day New Brunswick Avenue and New Market Road. The outnumbered Americans put up a strong defense by sniping at the enemy's flanks but were unable to stop the British column. General Leslie reported that several men were killed and wounded on both sides. Private Sullivan reported: "The 28th Battalion had one man killed and a captain and 15 men wounded. The enemy had four men killed in the wood, beside their wounded, which they carried off, except three that were taken prisoners."

Brigadier General Sir William Erskine recognized that it was difficult to directly engage the Patriots in wooded places where they knew the terrain and could obtain cover. He reasoned that an ambush might be a more effective way to deal with the wily yeoman farmers. On February 1 he sent a small party of foragers to Drake's farm near Metuchen. In the meantime, he kept a force of a thousand British and Hessian troops concealed behind a nearby hill. The ambushers were elite battalions of grenadiers and the crack 42nd Highlanders, all supported by eight cannons.

The 5th Virginia Regiment of the Continental Line, commanded by Brigadier General Charles Scott, promptly fell into the trap by pouncing on the small foraging party. Suddenly, Erskine's combined force appeared as if out of nowhere. The Redcoats were astonished when the Americans did not run, but instead chose to fight so fiercely that the much larger British force was driven back. The Virginians then broke through the grenadier battalion by launching a vicious bayonet charge. Heavy cannon fire slowed the American advance, but the Patriots continued fighting until the British withdrew and fled back to New Brunswick. The Crown forces suffered 136 killed or wounded. The Americans lost 30 to 40 men.[5]

The humiliating outcome of the foray near Samptown provoked the British to mount a major offensive. They returned on February 8 with more than 6,000 men, almost one quarter of all British troops in New Jersey. The expedition was ordered to attack Quibbletown and capture the supplies the Americans had collected and stored there. Now under the personal command of Lord Cornwallis, these troops fought their way into the village against heavy American resistance. Hessian Captain Johann Ewald describes this action from Raritan Landing, New Brunswick to Quibbletown:

The road leading from Raritan Landing to Quibbletown ran

continuously through the woods, in which three devastated plantations were situated. At the first plantation, I ran into an enemy post of riflemen who withdrew after stubborn resistance, of whom several were killed and captured on their retreat. We followed this party so swiftly that we arrived with them before Quibbletown at the same time. The place lies on two hills between which a creek that winds through a ravine that is spanned by two bridges. The stone walls around the gardens as well as the houses on both sides of the ravine were occupied by enemy riflemen who abandoned the village after strong resistance when artillery was brought up and withdrew into the wood on the other side of the village.[6]

The attack on Quibbletown was two pronged. Cornwallis ordered General James Grant to drive toward Samptown with two regiments to relieve pressure on the attackers. Grant had publicly denigrated the fighting skills of American soldiers and was generally despised by the Americans. Brigadier General Charles Scott, and a separate unit commanded by Brigadier General Nathaniel Warner, resisted bravely, but they were forced to surrender the entire store of provisions at the depot. The Americans fell back into the Watchung hills, wisely sidestepping possible annihilation by an enemy vastly superior in numbers and weapons.

As Lord Cornwallis' columns withdrew back to New Brunswick, the Patriots inflicted heavy losses on their flanks and rear. In many ways this retreat was similar to the British withdrawal from Lexington and Concord that had taken place two years earlier. Thirty British soldiers were killed and thirty were wounded during this engagement. The Americans had six killed, twenty wounded and six captured.[7]

The Battle of Spanktown & a Violent Encounter at Bonhamtown

British Colonel Charles Mawhood commanded an elite force of grenadiers and light infantry. Mawhood had shown superior leadership at the Battle of Princeton, and his regiment was regarded by the entire British army as a top unit. His objective was always to "surprise, surround and extirpate the Rebel army or at least a large piece of it."[8] He headed into harm's way on February 23, 1777 when he attacked Spanktown (Rahway), a ten-mile march from his headquarters in Perth Amboy. He had been told that the place was lightly defended.

As Mawhood's special forces approached the town, they met a small detachment of New Jersey Militia herding cattle and sheep along St. Georges Avenue in the vicinity of Robinson's Branch of the North Branch of the Rahway River. Mawhood ordered Captain John Peebles, who led the Grenadier Company of the 42nd Highlanders, to lead an attack on the Americans. As they rushed in, a force of William Maxwell's Continentals sprang up from a concealed position and fired a crashing volley of musket fire into the Redcoats. This ambush completely surprised the grenadiers, and they were slaughtered as they bravely tried to rally.[9]

In this intense engagement twenty-six grenadiers fell. The British fled back to Amboy, and the empty wagons they had brought to carry away captured supplies were instead loaded with their wounded men. The Redcoats retreated through what is today Woodbridge Township. Their path, now Route 35, was heavily wooded, which made it very difficult to pass through. Peebles estimated that the trek was twenty-eight to thirty miles long, when it was actually less than ten miles. The Americans took full advantage of the local terrain by constantly sniping at Mawhood's columns from all sides as they fled down the rough trail. By the time they reached Amboy, Mawhood had lost between seventy-five and one hundred men. American losses at the Battle of Spanktown were five killed and nine wounded.

Two weeks later, on March 8, the desperate Redcoats once again took the offensive. On that day they encountered what they later characterized as a bunch of "American hornets."[10] The Americans were growing stronger and were retaliating by making incursions with large forces deep into British-held territory, a belt stretching from New Brunswick to Amboy. On April 10, a Redcoat detachment on the front lines was approached at Bonhamtown, now Edison, by Maxwell's brigade of 2,000 troops. The best account of this clash comes from British officer Archibald Robertson, who served as a Lieutenant General in the Royal Engineers:

> *Between four and five in the evening the rebels, in a body of about 2000 under Brigadier Generals Stephens and Maxwell, attacked Piscataway, where the 42nd regiment is cantoned, who beat them back nearly three miles to their camp near the heights near Metuchen, with the loss of six privates and 3 sergeants killed, 2 officers, 2 sergeants and 15 privates wounded. The Rebels lost a great number and had a captain and 21 taken by the 42nd and 11 taken by the light infantry who had a sergeant wounded.*[11]

This incessant and largely triumphant combat increased the confidence and expertise of both the militia troops and the Continentals. British Lieutenant General Sir Charles Stuart wrote, "The rebel soldiers, from being accustomed to peril in their skirmishes, begin to have more confidence. The wounding and killing of many of our rear guards, gives them the notion of victory and habituates them to the profession."[12]

Over the next month the Raritan Valley and the Millstone areas were stripped by both the Redcoats and Patriots. After threshing, farmers had to hide their wheat under the straw in barns to preserve it from the hungry soldiers. In many cases, none could be saved for seed for the next planting. Cellars, houses, pigpens and hen roosts were all pilfered. Anything of value was stolen. The Continental Army did little to intervene while they were trying to survive in Morristown during the winter and spring of 1777. Widespread desertions and a smallpox epidemic there threatened to undo the gains at Trenton and Princeton.

Had Washington been a less cautious commander, he might have made the fatal error of confronting a superior force on the plains of

Somerset County. His restraint and deception during this time proved to be his greatest act of leadership as the American commander-in-chief. Since the weather began warming in May 1777, no major engagement had transpired since the Continental Army had driven the British through the streets of Trenton and Princeton in January of that year.

Chapter Four

The Bloody Battle of Bound Brook

Heroic Patriot Stand or Humiliating Defeat?

WITH HIS ARMY becoming less stable each day, Washington reconsidered his tactics and chose instead to conceal his army for the remainder of the winter of 1777 behind the Watchung ridges at Morristown, New Jersey. The distance from Morristown to the British winter base in New Brunswick was forty miles. A depressed General Washington wrote to John Hancock on January 7, "The Severity of the Season has made our Troops, especially the Militia, extremely impatient, and has reduced the number very considerably. Every day more or less leave us. Their complaints and the great fatigues they had undergone, induced me to come to this place [Morristown], as the best calculated of any in this quarter, to accommodate and refresh them.[1]

All military action during the early months of 1777 was on the front line outposts that stretched from Springfield south to Bound Brook and Millstone. The British in New Brunswick overestimated the strength and condition of Washington's Army, and remained on the defensive there until April. With the Patriot militia continuing to block their foraging raids, the frustrated British decided to retaliate with a large-scale attack on the main American outpost at Bound Brook. This assault would also afford an opportunity to forage the area for food, fodder and anything else of value along the way. The expedition was planned as a surprise. Preparations were carried out

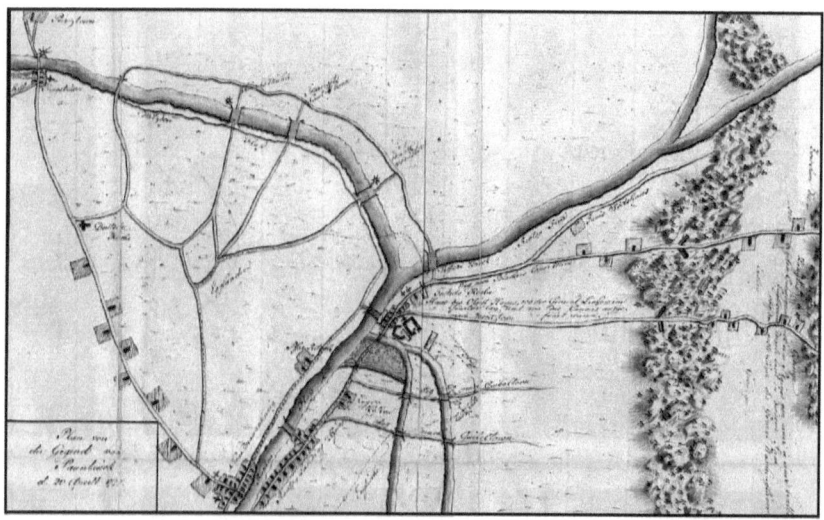

Map by Hessian Captain Johann Ewald depicting the Bound Brook area and the plan of attack. New Brunswick is at the bottom, and the Bound Brook outpost at the center. *(Harvey A. Andruss Library, Special Collections, Bloomsburg University of Pennsylvania)*

with such secrecy that most of the British troops in New Brunswick and sympathetic American informers did not learn of the plan until after the columns had departed on the offensive in the early morning hours of April 13, 1777.

The Bound Brook outpost was eight miles up the Raritan River northwest of New Brunswick. It was well positioned to block the enemy from moving supplies up the Raritan River, and to protect the prosperous farms with their well-stocked barns along the Raritan and Millstone Rivers. The Americans patrolled three miles along the north riverbank, from the Van Veghten Bridge at present day Finderne Avenue in Bridgewater to a bend in the river at the Queens Bridge at South Bound Brook. From here they could closely observe enemy movements at Raritan Landing Bridge in New Brunswick. While it was considered the strongest post on the American front line, less than 1,000 men were stationed there. The mission of this detachment was to warn the Continental Army base in Morristown of any impending invasion, and to offer resistance to delay any attack.

The Bound Brook outpost was commanded by Major General Benjamin Lincoln. He led the 8th Pennsylvania Regiment and a detachment of the 4th Continental Artillery, along with some militia troops. They were supported by three cannons, all of which were three-pounders. The number of troops there had been reduced by expiring militia enlistments to about 500 men by mid-March. Only the Pennsylvania regiment and the artillery company remained. Major General Lincoln was concerned. His backup, the main American army, was fifteen miles away in Morristown. He reminded General Washington that he had limited defense capability if Bound Brook was attacked, and kept wagons ready in case a sudden evacuation was required.[2]

The British attack objective was to capture Bound Brook and its three bridges, all of which might then be used to mount an expedition against Washington's army at Morristown. Their strategy was to surprise and engage the American riverbank outpost, then encircle it to prevent the small garrison from escaping into the nearby Watchung Mountains. In the confusion that ensued, the British would also attempt to capture Major General Benjamin Lincoln.

Blunders on Both Sides as Bound Brook is Attacked

Four columns, totaling 4,000 Redcoats troops with their Hessian allies, marched the eight miles from New Brunswick to Bound Brook during the night and early morning of April 13, 1777. What is astonishing is that this large force went completely undetected by American sentries, and no warnings were issued by the many Patriot citizens living in the area between the two places. Most of the blame for this lack of alarm was placed on careless sentries who were posted along the Raritan River. A local Loyalist farmer was also suspected of learning the American password and passing it on to the enemy.[3]

The British force arrived outside the town before sunrise and rested until daybreak. The signal to launch the onslaught would be

Old Stone Bridge, Bound Brook, Site of Ewald engagement. *(Author's collection)*

at the moment they heard the American sentries shouting "All is well," followed by a morning gunshot that reconfirmed that all was secure. These scheduled signals were heard at dawn by the stealthy-approaching Crown forces.

British General Cornwallis, second in command at New Brunswick, had previously ordered Hessian Jaeger Captain Ewald to draft a plan of attack on Bound Brook. Ewald writes, "Lord Cornwallis showed his confidence in me by with entrusting me with drawing up a plan for a surprise attack on Bound Brook. But since it was necessary for a column to cross the Raritan above Bound Brook, the attack was postponed until spring."[4]

The attack began when Captain Ewald's small unit of Hessian Jaegers was ordered to advance to the town before dawn to gather intelligence about the strength of the Patriot defenses. The Hessians, German mercenaries who comprised a select special force, were feared by their American opponents because they were recognized as the deadliest light infantry in the world at the time. They carried short carbine-type German hunting rifles and were clad in green and brown to blend in with their surroundings, in contrast to the blazing red wool uniforms of their British allies. They also wore tall helmets that tended to exaggerate their height and make them look more fearsome.

Ewald's band of thirty Jaegers was the first to reach the town. He reported, "I was ordered to form the advanced guard of General Grant's column. At daybreak I came upon an enemy picket on this side of the stone causeway which led to Bound Brook through a marsh...."[5] At this point Ewald had to cross over a stone-arched bridge that spanned Bound Brook on the edge of the town. A few yards from the end of the span the Patriots had built the "Half Moon Battery." This redoubt was the fortified position which blocked entry into Bound Brook.

Ewald approached the riverbank hamlet with his company. His orders were not to actively engage the Americans but instead to stay under cover to assess the enemy defense positions and then return with vital information. Instead, when the captain reached the bridge at the north end of town, he fearlessly ordered his men to make a frontal assault on the fortification guarding the town. They charged over the stone bridge directly into intense musket and cannon fire from the American position. As the Hessian Jaegers poured over the narrow span they were immediately pinned down by deadly fire from the redoubt.

Amid the turmoil on the bridge, the commander immediately sensed that something was dreadfully wrong. His rangers, combat-trained for both rugged terrain and urban fighting, were outnumbered and outgunned, and they were being slaughtered. Why had his small unit been ordered to lead a suicidal assault when thousands of troops of the main force would soon be arriving? The German troops were surrounded by hundreds of menacing Patriot soldiers.

The resounding noise of the intense musket fire exchanged during the disastrous charge of the Jaeger company alerted all 500 Patriot troops defending Bound Brook. Meanwhile, after about fifteen minutes, the valiant assault by the Jaegers had faltered. They were close to being annihilated. They narrowly escaped this fate when

Colonel Donop's troops of the main column appeared. They had been following close behind while advancing north from New Brunswick along present-day Easton Avenue. The fresh reinforcements crossed the Queens Bridge over the Raritan River a few yards from the redoubt and quickly overwhelmed the Half Moon Battery's position. The Patriot troops were forced to quickly abandon their positions. Ewald's survivors, together with the thousands of troops from Donop's column, skirmished with the outnumbered Americans who stubbornly resisted their advance through the streets of Bound Brook. Ewald described the action on the stone bridge in his diary:

> *At daybreak I came upon an empty picket on this side of the stone causeway which led to Bound Brook through a marsh along the Raritan River for five or six hundred paces over two bridges. The picket received us spiritedly and withdrew under a steady fire. I tried to keep as close as possible to the enemy to get across the causeway into the town at the same time. This succeeded to the extent that I arrived at the second bridge at a distance of a hundred paces from the redoubt which covered it and the flying bridge (Queen's Bridge). The day dawned and I was exposed to murderous fire.*[6]

Ewald added, "Luckily for us Colonel Donop's column appeared after a lapse of eight or ten minutes whereupon the Americans abandoned the redoubt. We arrived in town amidst a hard running fight. The greater part of the garrison of the redoubt were either cut down or captured."[7]

Ewald losses were due to his having been confused about his orders. His British commander had been vague about the mission and did not speak German. The Jaegers had been directed only to gather intelligence or possibly divert attention from the main forces by a feint, not an actual attack. The valiant German captain was later admonished by General Howe for being too aggressive, an act which alerted the Patriot troops that a major British offensive had begun.[8]

The main British forces approached from other directions minutes after the Jaegers made their fatal charge. The battle soon turned against the outnumbered Americans. Now realizing that they were outnumbered, they swiftly fled back toward the mountains.

The fighting at Bound Brook had lasted only ninety minutes. The Battle of Bound Brook has been largely overlooked by historians on both sides. This action was not decisive, so the engagement has been variously interpreted as either a humiliating rout or a courageous stand. In general, American military observers often neglect setbacks, but amplify triumphs.

The old stone-arched bridge where the Jaegers were pinned down still stands today. Built in 1731, it is one of the few existing battlefield resources in New Jersey where a first-hand account of the action that took place there exists. The triple arch crossing was constructed as a link over the old channel of the Bound Brook. It was part of Old York Road, one of the main highways across New Jersey that connected New York and Philadelphia in the 18th century. When a nearby railroad was built the Bound Brook was diverted so it would no longer pass under this bridge. Since the early 1870s the Old Stone Arch Bridge has been almost completely buried by fill due to construction of the railroad embankment. It lies exposed above the tops of the arches and underlies a more recent road. Sadly, eighteen-wheeler trucks regularly pass over the remains of this fragile structure. Its location is off South Main Street at the approach to an industrial zone. It was placed on the National Register of Historic Places in 2008.

Stunned Patriots Spring into Action

The predawn surprise strike at Bound Brook was a four-pronged assault. English Colonel William Harcourt broke off his column of 2,000 men from the main assault as it advanced toward Bound Brook. They moved up through today's Franklin and Piscataway Townships to Weston Canal Road and crossed the river on the Van Veghten Bridge onto Finderne Avenue in Bridgewater. The objective of this thrust was to capture General Benjamin Lincoln. This position was defended by Proctor's American artillery with their three cannons.

Hessian Colonel von Donop, along with the main body of his troops, continued up the left or Bound Brook side of the river. To block a possible the American route of retreat to Morristown a second column, commanded by Major General James Grant, broke off and crossed the bridge at Raritan Landing. They advanced up River Road on the right side of the river toward Bound Brook. They avoided main roads and moved as quietly as possible.

A third column of light infantry commanded by Major Maitland swept along to the east through Piscataway and moved along the base of the Watchung Ridge to cut off any escape by the Americans to the hills in the north. These columns reached positions surrounding the town near daybreak in an attempt to encircle the town and block an American retreat. Major General Lincoln's headquarters was nearby in the home of Philip Van Horne. A short distance south of this house the Patriots built a blockhouse and surrounded it with earthworks to prevent any river crossing from that direction. Soon the tranquil countryside was shattered by volleys of cannon blasts and clattering musket fire.

Sentries near General Lincoln's headquarters at the Van Horne House (now on East Main Street in today's town of Bridgewater) first heard drums and cannon fire at daybreak. The sudden appearance of mounted Redcoat grenadiers surprised Americans who cried out "To arms!" The frantic shouts alerted Lincoln and his staff, who were terrified to see the enemy troops a mere 200 yards away. Lincoln jumped out a rear window, vaulted onto his horse and began leading his surviving soldiers out of the town through a gap in the closing British pincers. In the ensuing confusion one of his aides was captured, along with all the general's baggage and records. Colonel Von Donop reported that General Lincoln "must have retired en *Profond Négligé.*" (without all of his clothing).[9] Lieutenant Simon Spaulding of an independent company mustered from the Wyoming Valley of Pennsylvania is credited for alerting American troops and allowing the majority of them to escape.

It should be mentioned that local historians disagree on exactly where General Lincoln was that morning. While most believe that he was in the Van Horne House, others claim that he was in the house of Peter Williamson, known as the Battery house, situated at the eastern end of Lincoln Avenue (now Bound Brook's main street). That site was also the location of the Half Moon Battery, where the Patriot defenders fired into Ewald's men crossing the stone bridge. Captain Ewald describes the action at the Van Horne House: "In the evening Colonel Harcourt with 50 horse, two light infantry battalions and a battalion of British grenadiers crossed the Raritan River below the Van Veghten bridge and arrived behind Horne's plantation where generals Lincoln and Wayne lay in their quarters under cover of the three pounders mounted in the rear of the enemy quarters. The guard was partly cut down and partly captured, the three cannon seized and the two generals fled without their breeches.[10]

The attack over the Van Veghten Bridge by the Cornwallis column drove the Americans from the blockhouse and decimated Proctor's artillery company. Most of its men were killed or captured and their cannons were seized. Had the alarm from the sentries come only a few minutes later or had the Cornwallis plan for surrounding the Americans been better organized, the entire American force at Bound Brook might have been encircled and captured. However, the British plan failed. Ewald's premature arrival alerted the Americans, and the fourth column led by Major Maitland, closing in from the north near Green Brook, arrived too late. The American force at Bound Brook was not surrounded, and the survivors of the 500-man garrison eluded the thousands of Redcoat invaders and headed for the nearby hills.

The escaping Americans swiftly fell back along Vosseller Avenue toward the first ridge of the Watchung Mountains. They managed to maintain brisk musket fire as they retreated, and after reaching the higher ground they rallied and attempted to make a stand. But overwhelming numbers of British forces began reforming and returning heavy fire, so the Patriots continued retreating into the hills.

The British looted the outpost at Bound Brook and returned to New Brunswick later that morning. They took along Proctor's three cannons, as well as ammunition and supplies. Ewald's diary stated that "afterward the place was ransacked and plundered because all the inhabitants were rebellious-minded, and then the entire corps withdrew along the road from Bound Brook, New Brunswick—and the enemy who had rushed support from Basking Ridge showed himself only at a distance."[11]

A large force under Major General Nathanael Greene had been sent from Basking Ridge later in the day to reoccupy Bound Brook. The British had already left by the time they arrived, and Greene sent a detachment to harass their rear guard. They caught up with the British near Raritan Landing, where they killed eight of the enemy.

American commanders were shocked by the loss of Bound Brook. General Washington immediately responded by reducing the number of outposts in Somerset County to prevent others from also being taken. Curiously, the British did not follow up their victory. They choose not to occupy Bound Brook or pursue the Patriots. They had failed to capture the Americans and General Lincoln escaped to fight another day. Most then town's surviving residents had fled with the American soldiers and were hiding in the Watchung hills.

This clash at the riverbank outpost had a profound effect on the course of the entire war. General Washington, now with only 4,000 ragged men remaining in the entire Continental Army, left the secure encampment at Morristown on April 19, 1777 to move to the front lines on the high ground of the Middlebrook Encampment. The defeat at Bound Brook proved that there was a real possibility that the British Army and its Hessian allies could sweep across central New Jersey. By doing so they could take the American capital at Philadelphia and end the American rebellion.

Aftermath of the Battle of Bound Brook

The extent of losses at Bound Brook depended upon who reported them. General Howe claimed that about thirty Americans were killed and eighty to ninety were captured, while General Lincoln reported that sixty of his men were killed or wounded.12 Howe claimed no deaths, and only seven wounded among the British and Hessians.[13] Washington reported, "The enemy lost the post at Eleven O'Clock the same day, & our people took possession of it again," and that the Continental Army's losses were "trifling and not worth mentioning." He did, however, admit that between thirty-five and forty were killed or captured, and that three field cannons were lost.[14]

In a report to the Board of War, Washington admitted the capture of two cannons, two officers and twenty men from Colonel Proctor's Regiment.[15] General Greene reported to his wife, "The British Generals breakfasted and I [dined] at the same house that day."16 An interesting and vastly different perspective of the Battle of Bound Brook can be found in newspaper accounts that appeared within a few days of the action. The British *New York Gazette*, gave this account on April 27, 1777, eight days after the battle:

> On Saturday, the 12th Instant, Lord Cornwallis with the Generals Grant and Matthews, with a body of British troops, and Col Donop, with a detachment of Hessians, surprised a large body of Rebels at Bound Brook, about seven miles from [New] Brunswick, under the command of one Benjamin Lincoln, late secretary to the Conventions and Congresses of Massachusetts Bay, and a forward person in all the rebellious Proceedings of that Colony. The troops lay upon their arms until Day-break, and commenced the attack upon the rear of the Rebel quarters, who made so weak a resistance as only to wound slightly four of the soldiers. Above 100 of the Rebels were killed. Eighty-five were taken prisoners among whom was a fellow who passed for Lincoln's Aide-de-camp and two others under the style of Officers. The rebels taken

were brought to town [New York] in the beginning of the week. And are the most miserable creatures that ever bore the name of Soldiers, covered with nothng but rags and vermin. Three brass field pieces, musquets, ammunition, camp Equipage, papers, several horses, near Two hundred head of cattle, with sheep, hogs, flour, bread, etc. Were chiefly brought away, and the rest, such as Rum and salted provisions being very bad were destroyed.

The American view was published in the *Pennsylvania Evening Post* on April 15, just two days after the battle:

The enemy came out yesterday morning, as reported in a letter from Headquarters, dated April 14th, 1777 from [New] Brunswick, with an intent to surprise General Lincoln at Bound Brook and had like to effected their design by the carelessness of a militia guard upon one of the fords on the Raritan, but the General got notice of their approach time enough to withdraw himself and most of his men to a mountain just in the rear of the town. Our chief and almost only loss was our two pieces of artillery and with them Lieutenants [Charles] Turnbull and [William] Ferguson, with about twenty men of Colonel Proctor's regiment. A party of horse was pushed so suddenly upon them that they could not possibly get off. The enemy staid about an hour and a half, and then went back to [New] Brunswick. General Lincoln took his post again with a reinforcement.

Washington grew more apprehensive after the attack on Bound Brook. He believed that the incident could signal the beginning of the campaign season in New Jersey. He knew that his troops were not prepared to repel major British incursions. There were no further moves over the next two weeks, but the Americans soon learned that the enemy was preparing to again take the field on June 1, 1777.[17]

However, after reviewing the documents captured at Bound Brook, the British realized that the Americans were expecting them to cross the state to capture Philadelphia. Once again overestimating the American strength, they feared that the enemy would suddenly swoop down from the mountains and strike their stung-out column as it marched west toward the rebel capitol. They abandoned this theory, and instead believed that the American rebellion could be ended by drawing the Continental Army out of the mountains to fight a major battle on the plains of Somerset County.

The unfortunate surprise at Bound Brook alerted General Washington to the fact that the river post was of great strategic importance. He considered constructing strong fortifications there and assigning a large part of the Continental Army to defend it. General Greene surveyed the area and concluded that it was exposed on all sides to attack and would be difficult to defend against an enemy incursion. As a result, all remaining troops were withdrawn from the town and were soon joined by Washington's main army at the First Middlebrook encampment on May 28, 1777.

The action at Bound Brook was a pivotal event in the history of the Revolutionary War. This intriguing but overlooked event in New Jersey history is worth reexamining today. Two remarkable circumstances enhance the appreciation of the battle: a fresh perspective of the fray is given by the discovery of the intrepid Hessian Captain Ewald's diary, and the physical features of the battle site still exist and can be easily identified today.

On East Main Street a 4,000-pound boulder was unearthed by an Irish laborer working on the construction of the Delaware and Raritan Canal between the years 1831-1843. For many years it remained on the sandy bank of the canal. On April 13, 1897, the anniversary of the battle, it was erected on the site of the blockhouse that defended the road approaching the Queen's Bridge. Due to traffic congestion, it was later moved to the grounds of the old borough hall. Finally, to shift it closer to the battle site, it was erected in 1957 on the lawn of the Pillar of Fire building, where it can still be seen today.

When you arrive in Bound Brook you are in the midst of the

waterways, mountains and other landmarks that determined the flow of military action in 1777. Moreover, houses and outbuildings that played a role in the clash have been faithfully restored by devoted residents of the area.

Chapter Five

The First Middlebrook Encampment

May to July 1777

GENERAL WASHINGTON HAD previously ordered troops to guard the main gaps through the southern section of the first Watchung Ridge as early as the fall of 1776. These gaps were at Middlebrook, (where Chimney Rock Road cuts through the ridge today), the Quibbletown Gap along Warrenville Road, and the passes at Stony Brook in Watchung and at Scotch Plains (where Bonnie Burn Road now ascends the first ridge). The battles at Bound Brook and the Short Hills, in the spring of 1777, were attempts by the British to penetrate American defenses by moving through these passes to get behind the ridge and then occupy the interior of the state. But the Crown forces were never successful in penetrating the Watchung ridges.

The Continental Army moved to its first major winter encampment at Morristown, New Jersey in December 1776. Unlike the previous miserable year at Valley Forge, the winter was remarkably mild and food supplies remained adequate. In the 18th century, most military activity in the western world came to a standstill during the winter months. It now seems unusual that operations would be suspended for an entire season, regardless of the army's location or level of activity. Harsher weather in those days made winter encampments necessary, and time was needed for soldiers to rest and repair equipment. Washington learned the value of this strategy early in the war. In December 1775, half of his fledgling army had been annihilated during a raging snowstorm while trying to storm the bastion of Quebec City. In fact, the entire Canadian campaign had been a weather-related disaster for the American forces, and it provided a lesson which they never forgot.[1]

Hostilities typically broke off in the late fall, when both sides would go into permanent winter camp until spring. A secure place was selected, cabins were built, and many officers and men were allowed to return home on furlough. There were exceptions to the hiatus during the winter months. For example, Washington famously crossed the Delaware River on Christmas night to surprise the Hessians at Trenton, and then followed up with a victory at Princeton. When the fighting broke off in January 1777, the triumphant Patriots immediately headed to Morristown, New Jersey, where they remained camped until the end of May.

Historians persistently laud the victories of the Continental Army but neglect the fact that for the majority of the time the American soldiers were encamped. The most notable American long-term winter encampments were at Morristown, Pompton and Middlebrook, New Jersey, New Windsor, New York and Valley Forge, Pennsylvania. The British forces, on the other hand, had a comfortable and festive winter in the captured rebel capital of Philadelphia in 1777-1778, while their main army spent most of the other winters of the war secure in New York City.

Fortunately for the Americans, Washington's leadership on the battlefield and his growing popularity throughout the country helped attract new army recruits during the first winter encampment at Morristown. The entire army had shrunk to about 4,000 men in the early days there. To attract more recruits, any man enlisting for a period of three years received both a cash bonus and a land bounty warrant for 100 acres. Discipline quickly improved when the maximum punishment for soldiers rose from 39 to 100 lashes.

After the Battle of Bound Brook General Washington ordered Nathanael Greene to scout the surrounding locale for a place where the entire Army would be secure, while at the same time remaining close to any possible enemy action. Greene began exploring the first

Watchung Ridge and the valley that lay behind it. He selected this valley along the Middle Brook for the army's encampment, a place now known as the Washington Valley.

The unfortunate clash at Bound Brook was a wakeup call for General Washington. In late May 1777 he moved the entire Continental Army of 8,298 soldiers from Morristown twenty-five miles closer to the front lines. The troops gathered at Martinsville in Somerset County. Brigades set up camp along the Middle Brook south of the present center of the town.[2] Washington positioned his headquarters at a large marquee tent set up near near the Folcard Sebring house on Washington Valley Road in Martinsville.[3]

Washington's Army Advances to Middlebrook from Morristown

The encampment became known as Middlebrook. The presence of the American Army in this strategically located area would discourage the British from crossing New Jersey by land to take Philadelphia. It was at Middlebrook in the spring of 1777 that General Washington perfected his strategy of restraint by not engaging in a major battle when the odds were against him. This prudent approach eventually forced the British forces to leave the state.

Today the land encompassed by the First Middlebrook Encampment lies in the towns of Martinsville and Bridgewater. The entire army—including cavalry, artillery, and infantry—first pitched tents along the west branch of the Middle Brook. This stream flows through the Washington Valley in the towns of Warren and Martinsville. This area of lush farmland lies between the first and second ridges of the Watchung Mountains.

The army constructed earthworks and utilized old Native Lenni Lenape trails in an attempt to extend and increase its presence. Three redoubts were set up in the valley at Martinsville to guard against any attack through the Chimney Rock pass. These also guarded against incursions from the rear behind the first ridge from Pluckemin. One

MIDDLEBROOK

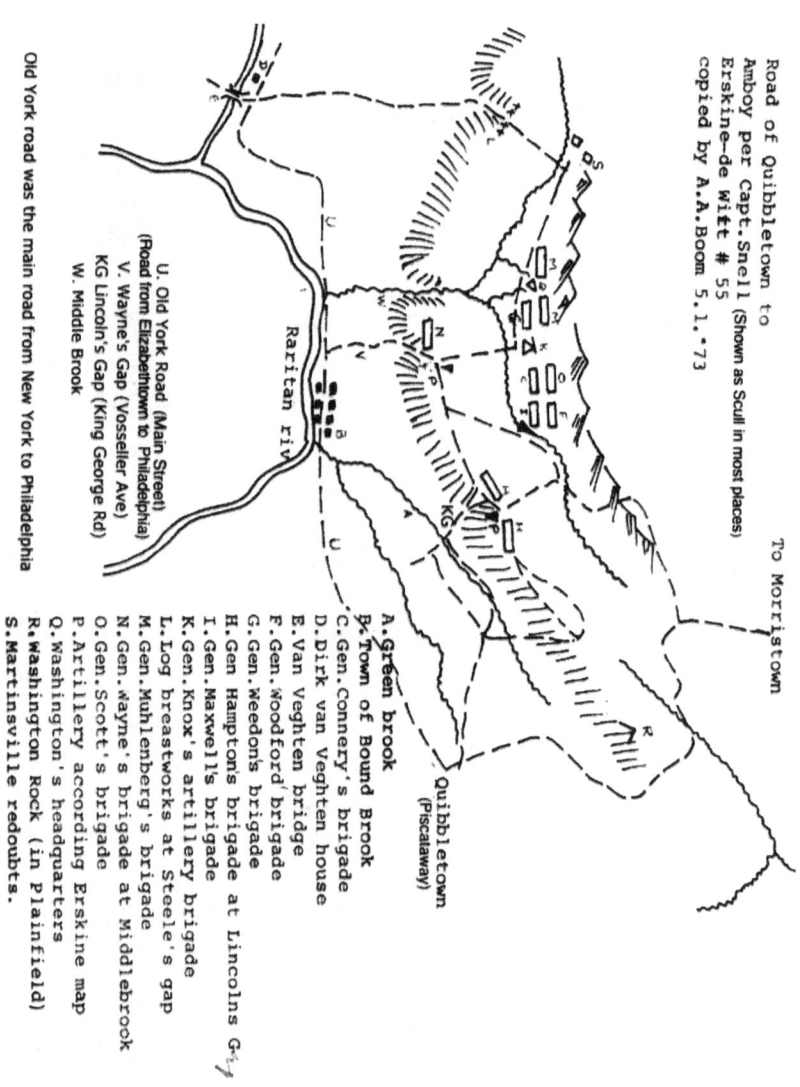

Erskind-De Witt Map 55, 1777, with additions by A. A. Boom in 1973. The Watchung Mountains rise across the upper left. The Middle Brook is the stream just west of Boundbrook on the left. Brunswick is just to the southeast. *(Courtesy New York Historical Society)*

of these large redoubts can still be seen today; signage directs visitors to these fortifications on Bolmer Farm Road.

The earthen redoubts built west of Chimney Rock Road and north of the Middle Brook stream protected the rear and right flank of the Middlebrook Encampment. A sign marking the place reads:

Local historian Ray Sanderson in Martinsville. *(Author's collection)*

Revolutionary War Redoubts, 1777 and 1778/79

When Washington's Army was encamped at Middlebrook in the spring of 1777, three earthen redoubts were built west of Chimney Rock Road and north of the west branch of the Middlebrook. They were 75 ft. square and 4 ft. deep and were equipped with a cannon and a complement of soldiers to defend the rear and right flank of the Middlebrook Encampment. The redoubts were also used in the winter of 1778-1779. This is the only remaining original redoubt in New Jersey.

The occupation of this critical site established a hold on the entire Watchung range. Wayne's position on top of the ridge held a commanding view of British positions on the lands below, stretching all the way to New Brunswick. The repositioning of the American Army along the Middle Brook and the adjoining high ground on the crest of the first ridge lifted the morale of the men in the ranks. New French uniforms and muskets soon began to arrive. Rigorous training and drilling ensured strict discipline. General Wayne wrote, "We are usefully employed in maneuvering. Our people are daily gaining Health, Spirits and Discipline—the spade & pick axe are thrown aside for British Rebels to pick up."[4]

The size of Washington's Army had grown from its low point of 4,000 to the current 12,000 soldiers. However, many of them soon had to leave to reinforce efforts to stop the advance of British General "Gentleman Johnny" Burgoyne in upstate New York. This still left Washington with 8,298 soldiers.[5] Included in this number were 2,660 sick or disabled men who were deemed unfit to fight. This army, consisted of both Continental troops and state militia soldiers, represented the entire Continental Army at that time.[6]

The British estimated that American strength was much greater. According to the notes of Howe's aide-de-camp, the Americans were being equipped with French-made Charleville muskets and clothing: "The few deserters who arrive now, all have French rifles. This confirms earlier speculations that the French have again sent large quantities of cannon, rifle, and clothing to the rebels. June 8, 1777."[7]

Washington had also gained confidence in the New Jersey Militia forces. These citizen soldiers, defending their homes and families, were now effectively routing the aggressive foraging British raiders. Washington was delighted when he learned that he could rely on them to successfully resist British foraging parties with little support from Continental troops. Militia units often went on the offense and aggressively pursued large numbers of British and Hessian scavengers. The general wrote,"The spirited manner in which the Militia of this State turned out upon the late Maneuvre of the Enemy has, in my opinion, given a greater shock to the Enemy than any Event which has happened in the course of this dispute...."[8] Milder weather soon set in, and when combined with contributions of hard currency from nearby towns, enlistments started were on the upswing.

General Washington kept his headquarters in the large marquis tent in the valley near the center of Martinsville for seventeen days. A sketch made by Charles Willson Peale during Washington's brief stay clearly shows the marquis tent. On June 12, British forces advanced to within four miles of the First Watchung Mountain and occupied the Raritan River Valley. Washington ascended what is today Vossellor Avenue and had his aides move his tent to high ground on the crest of the first ridge.

This was the campground of General Wayne's Pennsylvania Brigade. It appears that General Washington lived in this marquis tent for the entire seven weeks of the First Middlebrook Encampment. On June 29, 1777, he wrote to his brother John Augustine Washington from Middlebrook about British forces under General William Howe in the area, "I began to collect mine at this place; (a strong piece of ground) ten Miles distant from him [Wayne's camp along the crest of the first Watchung ridge], where I have now been in my Tent about 5 Weeks." This headquarters tent is now in the collection of the Museum of the American Revolution in Philadelphia.

A rock ledge lookout on the campground provided a commanding view of the British positions and enemy movements for an area stretching ten miles, from Bound Brook to New Brunswick. Wayne's Brigade built earthworks along old Native paths in the valley and hillside. The fortified camp gave the Americans an established hold on the first Watchung ridge that stretched north twenty miles to Springfield. Washington's Continental Army remained there from May 28 until July 2, 1777.

When the American Army pitched camp on the heights of Middlebrook, the main road over the first ridge was called "the Mountain Road," but it was also referred to as "The Basking Ridge Road." This is present-day Vossellor Avenue. It ran from Old York Road in Bound Brook over the first Watchung ridge through the pass known as Wayne's Gap. It then descended down through Martinsville and crossed the Middle Brook in the Washington Valley. Finally, it crossed the Washington Valley Road and climbed Mount Horeb Road toward Basking Ridge. Vossellor Avenue was the major road that bisected the Middlebrook Encampment area.

The Middlebrook Campground area was well known to General Washington. He passed through it early 1777 after the Battle of Princeton, on his way to Morristown. His army entered the pass via today's Chimney Rock Road and emerged in the Washington Valley behind the first ridge.

Middlebrook had obvious logistical, topographical, and geographical military advantages. The high ground made it an easily defended natural fortress. In those days, when men and horses arduously hauled heavy cannons and supplies, its elevation made it practically impregnable. The site had a 30-mile panoramic view of the no man's land that stretched over the plains of central New Jersey, extending east and southeast of Elizabeth to the Amboys and south to British-held New Brunswick. The entire population of the surrounding Somerset County was only about 10,000 residents at the time. While it could not support the entire army, the agriculturally-rich country could contribute to ease the difficulties produced by an insufficient food supply. The populace was generally patriotic and receptive to providing labor to the American army, and the dependable New Jersey Militia could always provide military support. Finally, Middlebrook had a plentiful water supply and abundant trees for the construction of huts and provision of firewood. The Continental Army at Middlebrook was in a position to block British overland expeditions, and if overwhelmed it could live to fight another day by hastily retreating into the safety of the hills of Morris County and even to the mountains of Sussex County farther west.

Before implementing any of their strategies the British had to provoke the American Army into moving down from the high ground to the open plains below. Here they could engage them in a conclusive European-style battle fought with linear tactics. Confronting a vastly superior foe would mean certain defeat for the Patriot forces, who totaled less than one half the size of their opponents. General William Howe, the top British commander in America, viewed the move to Middlebrook by Washington's forces as the first step toward achieving his own master plan. It placed the American forces in a position from which they could tentatively be lured down to confront his forces in a decisive battle.

The debilitated American Army had to remain intact so that the war could continue. A British victory here would end the war.

Setting up Campsites

The 8,000-plus men of the American force at the First Middlebrook Encampment represented forty-three regiments from five divisions. The troops were deployed to control all major roads into the encampment. Major General Nathanael Greene's 1st Pennsylvania Division, commanded by Brigadier Generals Muhlenberg and Weedon, were in the valley along the Middle Brook itself. With them was Lord Stirling, who commanded the 2nd Pennsylvania Brigade, William Maxwell's New Jersey Brigade and the Virginia Line were led by Generals Stevens, Woodford and Scott.[9] Benjamin Lincoln's Pennsylvania Division, with Anthony Wayne, were positioned at the crest of the ridge.

A map entitled "Road from Quibbletown to Amboy," drawn by Captain William Scull under the direction of Washington's chief cartographer Robert Erskine, provides a remarkably detailed depiction of the 1777 Middlebrook Encampment. It shows the placement of each regiment in the valley between the east and west branches of the Middle Brook, and American troop sites along the east branch of the Middle Brook and on the first Watchung ridge. The position on the ridge designated as "Wayne's B" lies along today's Vosseller Avenue and across the road onto Hillcrest Road. The map was likely drawn about May 28, 1777, the day Washington arrived.[10]

Another map that depicts the layout of the Middlebrook Encampment is attributed to the well-known British officer Major John Andre. It has the words "here rebels lay in 77" written across the valley of the east and west branches of the Middle Brook between the First and Second Watchung Mountains.[11] Both maps verify the location of Wayne's Pennsylvania Brigade. Additional information to substantiate the site appears in General Washington's Orders and the orderly books of regimental commanders.

The encampment covered a large area. Its western boundary ran for over two miles, along the west branch of the Middle Brook to

Dock Watch Hollow. The approach to the area was through present Washington Valley Road and was protected by the three redoubts on Bolmer Farm Road in Martinsville. The area around Martinsville was settled at the time of the Revolution, but the name of the town is not mentioned in any contemporary documents since it did not become official until 1827. All of Washington's letters and orders at this time identify the place as "Headquarters Middlebrook."

Brigadier General Anthony Wayne's command consisting of 1,400 troops, all spread along the crest of the first ridge. Orders for May 28 describe the alarm if the British attacked: "Three cannon would be fired in succession in front of General Wayne's Brigade in the Gap in the Mountain."

Wayne mentioned the site in a letter to Doctor Benjamin Rush on June 2. He notes his address as "Camp at Mount Prospect." Other correspondence between Wayne and his commanders also refers to Camp Mount Prospect. This evidence clearly labels the site as the "Mount Prospect Camp." The adjacent pass was referred to as both Wayne's Gap and Lincoln's Gap during the war, and for the ensuing thirty years.[12] The camp covered all the land between the crest of the ridge and Miller Lane and extended one mile from Vosseller Avenue to Chimney Rock.

Although no documentary evidence has ever been found to describe the exact layout of Wayne's encampment, it likely faced out toward the scenic view that it commanded, and consisted of rows of enlisted men's tents. It also had a parade ground that separated the soldiers from their officers, as well as butcher's tents and kitchens. Latrines were normally situated about fifty yards from the main camp for sanitary reasons. General Lincoln's Brigade was ordered "to establish a grand parade for his division, to assemble his guards at, and to appoint field officers to visit the guard both day and night."[13] Although the precise location of this parade ground has never been firmly established, but it may have been along the flat ground of present-day Hillcrest Road. All documentation regarding the 1777 encampment suggests that the camp consisted of tents, and not cabins. Wayne's camp and the rest of the First Middlebrook Encampment site is described in detail by Rev. Dr. Abraham Messler in his history of Somerset County. His information was extracted from the letters of General Washington to Congress:

> *We may sufficiently indicate the precise place of the encampment by saying that it was on the right of the road leading through the mountain gorge in which Chimney Rock is situated, just where it rises up from the bed of the little stream and reaches the level of Washington Valley. A strong earthwork was thrown up about a quarter of a mile to the northwest, almost in the centre of the valley, as a protection to any movement approaching from Pluckemin. The whole of the pass leading through the narrow mountain valley was strongly guarded, while the brow overlooking the plain bristled with cannon. Just at the edge of the wood, east of Chimney Rock, huts were erected as quarters for the officers, and everything done which either safety or comfort demanded in the emergency. At Bound Brook a strong redoubt was constructed, commanding the bridge over that miry little stream, just north of the present railroad-crossing, looking to any attack to be made from the way of New Brunswick. Having taken, in this way, all possible precaution against surprise, he felt strong to abide the issue of events. The result justified his sagacity as a military tactician. ...On the apex of the Round Top, on the left of the gorge in which Chimney Rock stands, there are yet to be seen the rude remains of a hut which Washington sometimes frequented during those anxious months of 1777.*[14]

Life during the First Middlebrook Encampment

The first month at the new camp was an inactive but tense time. On June 14, after seventeen days in Martinsville, Washington moved his headquarters to the Wayne's Brigade campsite on the crest of the ridge. He used a vantage point atop a local rock formation to observe British troop movements. The lookout post soon became known as Washington Rock at Middlebrook; today, it lies adjacent to the Eagle's Nest Museum and the home of Herb Patullo.

Daily events at the camp are noted in the General Orders of early June. Each soldier was commanded to maintain two days rations and a supply of ammunition on hand in the event the expected invasion began. As the troops began arriving there was an outbreak of typhoid fever. General Greene believed it was caused by the soldier's heavy diet of fresh meat. He attempted to curb the spread of the sickness by issuing each man a half pint of vinegar each day. As additional precautions, wells were covered with boards and animals were slaughtered as far from the Camp as possible.

Washington complained to the President of Congress that desertions were exceeding the recruiting rate.[15] The desertion rate increased even more after General Howe offered a pardon to all Americans joining the British side:

By His Excellency Sir William Howe

Knight of the most honorable Order of the Bath. One of his Majesty's Commissioners for restoring peace to the Colonies. General and Commander in Chief of all his Majesty's Forces within the Colonies lying on the Atlantic Ocean from Nova Scotia to West Florida.

Proclamation
Whereas it has been represented to me, that many of his Majesty's [European and American] subjects are compelled by force .or otherwise indDuuuced, to bear arms in opposition to the Re-establishment of the Constitutional authority of Government in

America, and are discouraged from returning to their Allegiance by ill-founded doubts of the reception such tender of their duty may meet with: I therefore Declare and, do hereby promise and engage, that all persons bearing arms as aforesaid, who shall surrender themselves to any officer commanding any part of his Majesty's Forces, on, or before the first day of May next, shall be entitled to Pardon for all offenses heretofore committed against his Crown and dignity and their estates and effects be secured from forfeiture, or Confiscation. That every Non-commissioned Officer and Private man who shall come in with his Arms shall receive also full value for them. That the American-born subjects shall be permitted to enter into any of the Provincial Corps, in his Majesty's Service. Or to return Home as they think fit. And that the British or Irish born subjects shall be taken into his Majesty's Service, or conveyed to the place of their Nativity at their option.
Given under my hand, at Head Quarters in New York the 15th day of March. 1777. W. Howe

During the 1777 encampment the following standing general orders were issued each day regarding living conditions in camp:

General Orders Camp Middlebrook- May 31, 1777 to July 5, 1777

Guard post and fortifications erected outside of the main camp: "The brigades on the right and left, front and rear of the camp in the valley between the first and second ridges are to establish small guards of Subaltern. One Corporal and eight Privates, in all passes leading to the camp, in order to prevent soldiers from straggling and the country people from coming into camp." - General Orders,

Middlebrook May 31, 1777. [The lonely soldiers manning these guard posts were allowed a gill of rum per man each day (about 1/2 cup), but it was not provided until their work was over.][16]

Trash: *"Vaults should be dug in the rear of each encampment, as repositories for any kin fog fifth. They should be covered with green boughs and fresh earth should be thrown upon them each day or two."* Also, *"All dead horses, cattle or other carrion removed to a distance from the camp and buried under cover. The commissary General to have his slaughter house at least a mile to the rear of the camp"* - General Orders, Middlebrook June 2, 3, 4, 1777.

Water supply: *"The Brigadiers are to have the springs adjacent to their encampments well cleared and enlarged"* - General Orders, Middlebrook, June 3, 1777.

"The General likewise recommends it to the Brigadiers to have the springs sought and opened and barrels sunk in them. For more conveniently supplying the troops with water." - General Orders, Morristown July 5, 1777.

Camp Kitchens: *"Temporary ovens to each brigade, which by men who understand it can be erected in a few hours. Baked in these will be much more wholesome than the sodden cakes which are but too commonly used."* - General Orders, Morristown July 5, 1777.

Patriot spies in New Brunswick reported that wagons and flat-bottomed boats were being relocated from New York to the Raritan River. This seemed to indicate that the British forces were preparing to cross overland through New Jersey in order to capture Philadelphia.[17] Washington had warned General Nathanael Greene to alert General Benjamin Lincoln, who had fled into the hills with his men after the Battle of Bound Brook, of this movement.[18]

By mid-June, General Howe had assembled the British Army from New Brunswick and other nearby towns in the Raritan Valley to an area that is now Franklin Township. The British, with their overwhelming superiority in numbers, desperately wanted to end the military stalemate by enticing Washington's Army down from the security of the Middlebrook Encampment onto the open terrain at Millstone and South Plainfield.

The high ground at Middlebrook prevented Howe from making a frontal assault, but feigning an overland march west toward the Delaware River might deceive the Americans into coming down for a full scale battle, an action they would surely lose.

Patriot resistance had grown stronger in the spring of 1777, and the British were forced to strengthen their foraging parties to over 1,000 men. This standoff came to an end on June 12 when Howe went on the offensive by moving his main army of more than 18,000 soldiers from New Brunswick toward the Millstone River. Swarms of well-organized local militia harassed the British supply lines as they moved from New Brunswick.

During the evening on June 14 Washington learned that Howe had moved most of his army towards either Somerset Courthouse or Millstone, and had set up a line of defense from the Millstone River back to New Brunswick. As a result, Howe's troops now occupied all of the country south of the Raritan River. At daybreak on June 15 Washington immediately moved all of his army from the Washington Valley to Wayne's campsite at the crest of the first ridge. The general wrote to his brother John Augustine Washington, "Finding Genl Howe was Assembling his whole Force (excepting the necessary Garrisons for New York &ca) at Brunswick, in this State, I began to collect mine at this place; (a strong piece of ground) ten Miles distant from him, where I have now been (in my Tent) about 5 Weeks."[19]

The cautious Patriots failed to appear to engage the British and were sneered at by Loyalist civilians. Howe's aide, Friedrich Ernst von Muenchausen, commented "Washington is a devil of a fellow, he is back again, right in his old position, in the high fortified hills. By retreating he supposedly intended to lure us into the hills and beat us there."[20] Washington had also learned from a British deserter that the sole reason for the massive invasion was to entice him out of the mountains.[21]

While Howe was entrenching his army on the south shore of

the Raritan and Millstone Rivers the American forces continued reinforcing their fortifications on the hill. But they did not venture down to confront the Crown forces. The British spent the next eight days threatening and challenging the Americans, but Washington did not leave his mountain stronghold. To his consternation, Howe finally realized that Washington could not be lured down from his position on the heights, and to assault the dug-in American positions on high ground would result in heavy losses. On June 19 the frustrated British army withdrew back to New Brunswick and made an extensive modification of their strategy. They began moving north to Perth Amboy. From here they could easily evacuate to Staten Island. That strategic move soon revealed their intentions. A long sea and river route could then be taken to approach the Rebel stronghold of Philadelphia, rather than march across the interior of New Jersey.

A revealing map of the First Middlebrook Encampment was prepared in 2019 by John Madden, a former Bridgewater Planning Director. Madden refined another map drawn by A. A. Boom in 1974 which was itself based on the original Scull/Erskine map of 1777. Madden's map clearly shows the locations of the different brigade camps. He then combined it with information from the 1754 Hill Survey of Somerset County, which showed the network of roads existing at the time of the First Encampment.

Madden also added features from an 1850 Somerset Land map by J.W. Otley and an 1860 map of Bridgewater. Both these documents indicate the encampment and fortification sites. This data was then all overlaid on an 1888 topographical map which placed the site of each brigade camp on flat, dry areas between the tributaries of the Middle Brook. Madden commented: "You can still get a feeling of what it was like at the Encampment in 1777 when you look south from Newman's Lane toward the wooded hills of the First Watchung Ridge."

Chapter Six

The Nest of the Eagle

THE AMERICAN FORCES moved to high ground at Middlebrook in late spring of 1777. They were well rested after spending five months at winter camp in Morristown. The Cornelius Vermeule Plantation, home to over 1,000 militia in North Plainfield, had served to discourage incursions through the passes in the Watchung ridges during that time. British General Howe, Commander of the Crown forces in America, made several attempts to entice the Continental Army down to the open flat land of Samptown, now South Plainfield. But General Washington knew that of the Crown forces were superior. He chose not to leave his secure position and resisted the temptation to start a campaign to liberate New Jersey.

This cautious approach earned him the nickname "American Fabius." The label was taken from the Roman General Quintus Fabius Maximus, who was regarded as the father of guerrilla warfare. Fabius avoided decisive battles when he knew that his army was outnumbered. He won by harassing enemies in relentless small skirmishes, causing attrition, disruption to supply lines, and a general reduction of morale. Fabius was also called the "The Great Delayer" for not attacking the superior Carthaginian forces of Hannibal.

Washington's cautious strategy was soon recognized and admired by military leaders and politicians in both America and Europe. John Adams, serving in the Continental Congress, acknowledged the value of this tactic, writing to his Abigail "Our Fabius will be slow but sure."[1]

The London Annual Register commented, "These actions and the sudden recovery from the lowest state of weakness and distress, to become a formidable enemy in the field, raise the character of General Washington as a commander both in Europe and America and with his preceding and subsequent conduct serves to give a sanction to that appellation which is now pretty well applied to him, of the American Fabius." By retreating and yielding the general kept the beleaguered Continental Army intact at Middlebrook.

Surgeon James Thacher, writing from Middlebrook in 1778, depicted Washington's physical appearance as follows:

> *Remarkably tall, full six feet, erect and well proportioned. The strength and proportion of his joints and muscles, appear to be commensurate with the pre-eminent powers of his mind. The serenity of his countenance, and majestic gracefulness of his deportment, impart a strong impression of that dignity and grandeur, which are his peculiar characteristics, and no one can stand in his presence without feeling the ascendancy of his mind, and associating with his countenance the idea of wisdom, philanthropy, magnanimity and patriotism. There is a fine symmetry in the features of his face, indicative of a benign and dignified spirit. His nose is straight, and his eye inclined to blue. He wears his hair in a becoming cue, and from his forehead it is turned back and powdered in a manner which adds to the military air of his appearance. He displays a native gravity, but devoid of all appearance of ostentation.*

In this same year Thacher again wrote, "General Washington is now in the forty-seventh year of his age; he is a well-made man, rather large boned, and has a tolerably genteel address; his features are manly and bold, his eyes of a bluish cast and very lively; his hair a deep brown, his face rather long and marked with the small-pox; his complexion sunburnt and without much color, and his countenance sensible, composed and thoughtful; there is a remarkable air of dignity about him, with a striking degree of gracefulness."

Washington's quiet strength and dignified reserve were often mistaken for pomposity. He was on a serious mission and had little time to bond with close friends. He lost his temper and was a poor

Panoramic view from the campground on the first Watchung ridge. *(Author's collection)*

public speaker, and yet was a prolific writer. When he did speak he was blunt, and looked people squarely in the eye. When disturbed, the father of our country could let loose a torrent of curses that would make even his hardened soldiers blush.[2]

A large British force advanced to Piscataway on June 1, 1777, but it was repulsed by Colonel Oliver Spencer's New Jersey Regiment and forced to return to New Brunswick. Howe was not achieving any success with his incessant attacks and counter attacks. American guerillas sniping from the nearby wooded roadsides were wearing down the numbers and the morale of the British Army. Even worse was that his huge force could no longer be fed. Time had run out for the British. The Continental Army had to be decisively defeated or an evacuation from New Jersey would be necessary. The American Fabius had to be lured down from Middlebrook.

On June 13, Lord Howe divided his army into three parts. One remained at New Brunswick, another was to march at midnight to Millstone, and the third was to advance to Middlebush. Earth works were thrown up at these places. Washington could observe this mass invasion from his Middlebrook overlook. Cautiously, he moved the American forces down the slope in the rear of Bound Brook. From here he could either skirmish with the Redcoats from safe higher ground or set up defense positions. He realized that this British incursion was a defiant invitation to come down to fight. Howe halted his advance at Millstone. The town was halfway to the American lines, eight miles away at Middlebrook.

June 14 was an unusually eventful day in the Middlebrook area. Howe's army was digging in along the riverbanks and Washington was advancing closer to the enemy.[3]

The Stars and Stripes at Middlebrook

While no written evidence or primary documentation has ever been discovered to prove it, it is likely that the American flag was first flown at the Middlebrook Encampment. The Washington Campground Association has recently obtained an appropriation from the of New Jersey Legislature authorizing the expenditure of ten thousand dollars for the design and erection of a suitable monument at the Encampment at Middlebook to commemorate it as the place where the American flag was first promulgated to the army.

Two facts support this claim: On June 14, 1777, the second Continental Congress passed the Flag Resolution, which stated: "Resolved, That the flag of the thirteen United States be thirteen stripes, alternate red and white; that the union be thirteen stars, white in a blue field, representing a new constellation"; at the exact same time, and for eighteen days afterward, Washington and his army were at Camp Middlebrook. The inference, which in the judgement of the members of the association and others is considered factual, is that Washington must have used the design of the present American flag as the official flag of his country at Camp Middlebrook. The news of its adoption would have reached the camp by courier in a couple if days. It could not be otherwise. It is argued that before Washington broke camp on July 2, he must have flown the new and official emblem of his country's liberties. The fact that no tactual record of it has been located does not mitigate against the conclusion.

The flag would have been the 1777 version designed by Francis Hopkinson, which is generally believed to be the first. Hopkinson was a delegate to the Continental Congress from New Jersey, and a signer of the Declaration of Independence. Since no original examples of the very first flag has survived, no one can claim with certainty to know its exact appearance. It was reported to have had seven white stripes, six red stripe and thirteen stars. The so-called Betsy Ross Flag was more likely a flag used for celebrations of anniversaries of the nation's birthday. Her descendants claimed it was the first flag, but there is no primary evidence of that, and the first documented usage of this flag was not until 1792.

The Crown forces spent eight days in a futile attempt to entice the American Fabius down from his strategic position. They finally abandoned their plan and returned to New Brunswick on June 19 and set up camp on both sides of the Raritan River. Howe was tempted to head to Philadelphia but feared an attack on his strung-out columns of troops by the Continental Army. He had learned that it had been expanded by militia regiments, and as usual he greatly overestimated its size and strength.

In June 1777, General Washington wrote to his secretary and aide-de-camp Joseph Reed, "The spirited manner in which the Militia of this State turned out upon the late Maneuvre of the Enemy has, in my opinion, given a greater shock to the Enemy than any Event which has happened in the course of this dispute."[4] After an unsuccessful foraging raid a week later by a detachment of 700 men from New Brunswick, Baron Munchausen, aide-de-camp to General Howe wrote, "Washington is a devil of a fellow, he is back again, right in his old position, in the high fortified hills. By retreating he supposedly intended to lure us into the hills and beat us there."[5]

Washington reported his status to John Hancock on June 20: "Sir, As I informed you [previously] then the main body of the enemy has returned to Brunswick again, burning as they went several valuable dwelling houses. We have constantly light troops hovering around them...I am inclined to believe that General Howes's return, thus suddenly made, must have been in consequence of the information he received that the people were in and flying to arms to oppose him."[6]

The British forces paused for a day in New Brunswick, offering Washington's forces a final opportunity to counter-attack and engage in a major confrontation. Soon disappointed, Howe reluctantly concluded that the American forces had no intention of leaving their secure position in the hills, and that he could no longer sustain his men in New Brunswick by foraging. His master plan for 1777 had failed. He had not defeated the Americans on the battlefield, occupied New Jersey, or marched across the state to capture Philadelphia. Howe was badly in need of a new strategy.

On June 21, the Crown forces were observed destroying their fortifications in New Brunswick so that they could not be exploited by American forces. Three days later the British began heading east to the Amboys, twelve miles distant. From that seaport on Raritan Bay they could easily embark to Staten Island, and from there make an assault on Philadelphia by taking a water route. This maneuver might also present them with an opportunity to encircle the Continental Army by getting behind the first ridge by following the passes in Scotch Plains and Springfield.

Advanced parties began plundering and burning Patriot homes as the British army passed through the present-day towns of Highland Park and Edison. General Washington, observing the activity in the British camp, saw an opportunity to take the offense while the enemy was strung out and vulnerable. On June 22 he divided his forces into three detachments: General Greene would confront the enemy lead columns, General Lord Stirling would attack New Brunswick and attempt to block the enemy from retreating to New York by water, and Washington himself would remain in the high ground at Middlebrook with the majority of his army poised to move where it might be needed most.

General Greene sent riders to Washington and Stirling, calling for an immediate attack after his scouts had learned the British routes of march. Unfortunately, the three American detachments lost contact with one other, so the messages never made it through. The record book of American Quartermaster Sergeant Simon Griffin confirms that General Stirling held his men outside New Brunswick while awaiting the other units. Unfortunately, this delayed his attack until most of the enemy had already passed back through town and were on their way to Perth Amboy.[7]

Griffin described the skirmishing as the British departed:

June 22, 1777: This morning the enemy began to burn the houses and destroy all the things that they could not get away. Our People began to play them with their field pieces and then with some small arms fired. But not many more than 3 of the enemy were killed, with one shot from our people and one wounded.

We marched into Brunswick and crossed the [Raritan] River and made a halt on a hill above the River in which time part of the army belonging to Gen'l Green came up with the rear of the enemy and had a smart fire for some time making the enemy retreat from one breastwork to another for some time. I returned over the bridge with all the Brigade and lodged in the house close by the stone church. I drew provisions. We stayed all day at Brunswick.

As Griffin reported, General Stirling's men quickly seized the main bridge over the Raritan River at Raritan Landing. From the top of a hill above the south bank of the river they could see the enemy setting fire to houses and barns and plundering the area as they withdrew toward Perth Amboy.

The Americans at New Brunswick might well have inflicted heavy losses on the enemy if their communications had been better and the weather more cooperative. General Stirling's men pursued the retreating British Army down the main roads that led to Perth Amboy, today's Woodbridge Avenue and Route 440. Loyalist civilian Nicholis Cresswell reported that the retreating troops devastated the countryside. "All the country houses were in flames as far as we could see. The soldiers are so much enraged that [they] will set them on fire, in spite of all the officers can do to prevent it."

With his three brigades and supplementary New Jersey Militia units, General Greene chased the retreating British as they headed east toward the Amboys. His force of 1,200 men and two cannons attacked the rear of the Redcoat columns as they passed through Piscataway on June 22. These efforts failed to disrupt the British or prevent looting. Greene's forces also failed to link up with Stirling, and the Crown forces reached Perth Amboy the next day.[8] Stirling's Brigade rejoined Washington at Middlebrook. Washington later commented that if the Greene's assault had been successful, it would have been devastating to the British since they would have been trapped along the narrow roads leading through Edison and Metuchen.

On their week-long march from the Millstone area back to New Brunswick, and then on to Perth Amboy, the invading armies of British and Hessian troops grew weary in the Jersey heat and humidity. Tired and thirsty, the troops were burdened with heavy backpacks, muskets, rations, and ammunition, and were continually attacked by militia units concealed in the foliage along the dusty, narrow roads.

The opposing New Jersey militiamen were more comfortable in their loose-fitting linen hunting shirts and broad-brimmed hats. They knew the lay of the land well and they fought Indian style, striking the foe where they were weakest and then disappearing when the enemy regrouped or brought in artillery. These citizen soldiers were infuriated by the wanton acts of the British foraging parties that had been plundering their farms and villages all winter. And now they were burning the homes and outbuildings of their neighbors as they passed through Somerset and Middlesex Counties. As a result, thousands of angry militiamen rushed in from the countryside to assist in harassing the Redcoats.

While General Washington and his army remained securely nestled at Middlebrook, he became aware of a flaw in his defenses. The northern flank of the American forces, near the passes through the hills at Watchung, Scotch Plains and Springfield, needed stronger protection. This was in the critical Short Hills-Ash Swamp area in what is today South Plainfield and Scotch Plains. There the ground rises to the west of Oak Tree Road in Edison, reaching its highest point on the site of the present-day Plainfield Country Club in the area called the Short Hills. Appropriately named, they are low and inconspicuous when compared to the first ridge of the Watchung Mountains, which towers above them a few miles to the west.

Washington assigned William Alexander, also known as Lord Stirling, to this unprotected flank. Stirling one of his most trusted and experienced combat officers. This wealthy gentleman farmer from nearby Basking Ridge was heir to his title through his Scottish lineage. His detachment of about 2,000 men left Middlebrook on June 24 and advanced fifteen miles east along the foot of the first ridge, through what is today Green Brook Township. Stirling arrived the next day at the 1,200-acre Cornelius Vermeule Plantation in North Plainfield, the largest militia base in New Jersey.

The New Jersey Militia could fan out from this assembly place to anywhere in the central part of the state to protect inhabitants from the plundering bands of British and Hessian soldiers. The post was garrisoned by as many as 2,000 men. This made it half as large as Washington's entire Continental Army during the winter of 1776 and 1777. Stationed at the Vermeule Camp were the militia of Hunterdon, Morris and Sussex Counties, the First New Jersey Regiment, the First Somerset Regiment, under Colonel Frederick Frelinghuysen, and the First Middlesex Regiment under Colonel John Webster. General William Winds commanded the more than 1,200 men New Jersey Militia who were billeted at this stronghold.

This militia base covered ninety-five acres along the east bank of Green Brook, between what is now Clinton and West End Avenues in North Plainfield. Also known as the Blue Hills Post, it was built around the large fort that guarded the main road from Bound Brook through Quibbletown to Scotch Plains. That road is now Front Street in Plainfield. The post was directly below Washington Rock in Green

Battle of the Short Hills, June 26, 1777. A 19th century woodcut of a signal beacon on the First Watchung Ridge. (*National Archives and Records Administration*)

Brook. The site was situated amid sparsely-populated open farmland.

In addition to defending the main road, the Vermeule base was specifically situated to protect two vital passes through the first Watchung ridge. The Quibbletown Gap, now abandoned, paralleled the present Warrenville Road, and passed over the ridge near Washington Rock. The Stony Brook Gap, now Somerset Street in Watchung, led into the Washington Valley. American defenders at the Vermeule base could easily rush to block these passes, since both were within two miles of the camp. The Vermeule homestead was a large Dutch house on the corner of what is today Clinton Avenue and Greenbrook Road. A large Victorian-style house that serves as a museum now stands on the site.

Lord Stirling Dashes to Protect the Flank at Scotch Plains

Stirling left the post at Vermeule's farm in Plainfield on June 24 with a force of 1,798 soldiers and moved to the Short Hills-Ash Swamp area to protect the exposed left flank of the American Army. He set up his headquarters at a central location along Inman Avenue in Scotch Plains. On that same day Howe began ferrying troops out of the Amboys to Staten Island. Washington was puzzled by this maneuver.

What was Howe's next move? He wrote to Stirling. "Try to get one or more persons upon Staten Island to give information of any Number of the Enemy that might come over...."9

On June 25 American lookouts on the crest of the first ridge of the Watchung Mountain observed a welcome sight—the Crown forces were evacuating New Jersey by crossing over the narrow Arthur Kill to Staten Island. Surely, this was a sign that the entire British Army was abandoning the state. The elated General Washington called his top commanders together at the Drake house (now on West Front Street in Plainfield) to quickly plan a new strategy.10

The abandonment of central New Jersey by British forces would allow the Patriots to reoccupy that area and alleviate the suffering of the beleaguered farmers in that Tory-infested land. However, Washington remained apprehensive for a few hours, remaining reluctant to move the army out of the mountains. He finally yielded to his more zealous officers, who tactfully scorned him for his lack of aggressiveness.

On the day before he left the safety of Middlebrook, he wrote to Joseph Reed, expressing his concerns:"I cannot say that the move I am about to make towards Amboy accords altogether with my opinion, not that I am under any other apprehension than that of being obliged to lose ground again, which would indeed be no small misfortune as the spirit of our troops and the county is greatly revived (and I presume) the enemy's not a little depressed, by their late retrograde emotions." Contemporary British sources deny it, but this major withdrawal may simply have been a feint to deceive the wary Americans; perhaps it was even Howe's final attempt to lure them out of their mountain stronghold at Middlebrook.

The Americans were both delighted and relieved to see the withdrawal of the Crown forces to Staten Island. Washington jubilantly observed the evacuation from his perch on a large rock at Middlebrook at the crest of the First Watchung Mountain. He was convinced the enemy was leaving New Jersey permanently, so he allowed the militiamen to disband and return to their homes. He then confidently moved all American forces down to the plains at Samptown (South Plainfield) and Quibbletown (Piscataway) and fanned out his regiments in a ten-mile arc to defend New Jersey from an unlikely counterattack. The American lines on the plains of central New Jersey now extended from Quibbletown north to the Short Hills and Ash Swamp. This frontline would enable the Continental Army to attack the rear of Howe's retreating forces if there was an opportunity, or to escape it if they were attacked. Stirling's detachments continued to advance from the Vermeule base to the Scotch Plains area as Howe's army continued to board longboats for their crossing to Staten Island.

A Terrifying Surprise

On the evening of June 22, 1777 General Howe was amazed. He had just learned that Washington was finally leaving the security of his mountain stronghold. Finally, he thought, this unexpected move might provide the long-awaited opportunity to engage and conclusively defeat the Americans and end the rebellion. He immediately halted the evacuation to Staten Island and ordered his forces to about-face and head back into central New Jersey. As mentioned, some historians believe the withdrawal to Staten Island was a carefully orchestrated trick to draw Washington out of the hills. More likely, Howe had been alerted to Washington's imprudent move, and knew that the American left flank at Edison and Scotch Plains was unprotected. Stunned by Howe's directive, the dumbfounded Washington followed the enemy re-invasion by shifting his observation post. He shifted it five miles—from Middlebrook north to the iconic Washington Rock in Green Brook—a vantage point easily reached by following the ridge line of the mountain.

Charles Willson Peale, the renowned artist, was a captain in the Pennsylvania Militia at the time. He recalled, "We expected to move down towards Amboy, but early in the morning we understood that the Enemy were moving toward us."[11] Stirling's Brigade, vastly outnumbered and out gunned, stood alone. To defend the vulnerable flank and the passes through the first ridge of the Watchung Mountains, they formed a line of defense that stretched the two miles from Edison to the first ridge at Scotch Plains. The Battle of the Short Hills was about to begin.

Chapter Seven

The Battle of the Short Hills

The Defeat that Saved the Continental Army

BY JUNE 1777, British forces had occupied New Jersey for seven months. Repeated attempts had failed to lure the Continental army out of the hills for a major battle. British General Howe had finally accepted the fact that Washington was not willing to engage in a full-scale European-style battle on the plains of Quibbletown and Samptown, today's Piscataway and South Plainfield.

Howe also did not wish to risk a march across New Jersey to capture Philadelphia. He feared that his strung-out wagon convoys might be overwhelmed by Patriot marauders swooping down from the hills of the Middlebrook Encampment. The occupation of the Patriot capitol would surely break the back of the American rebellion. But to achieve this objective another strategy was needed. The capital city had to be reached by water.

The British Army left New Brunswick on June 22, 1777 and marched east to Perth Amboy, where it could make preparations to cross the Arthur Kill to Staten Island. From there a British fleet could transport the troops to the head of Chesapeake Bay. From the Head of Elk (now Elkton, Maryland) they could sail north eighty miles to seize Philadelphia. Washington was observing the enemy troop movements from his lookout post at Middlebrook. On June 24 he sent two detachments under Generals Greene and Stirling down from the hills in an unsuccessful attempt to strike the vulnerable British enemy as it left New Brunswick.

Sir William Howe was frustrated when he arrived at Perth Amboy with his army. His final attempt to lure Washington down from the

The Battle of the Short Hills

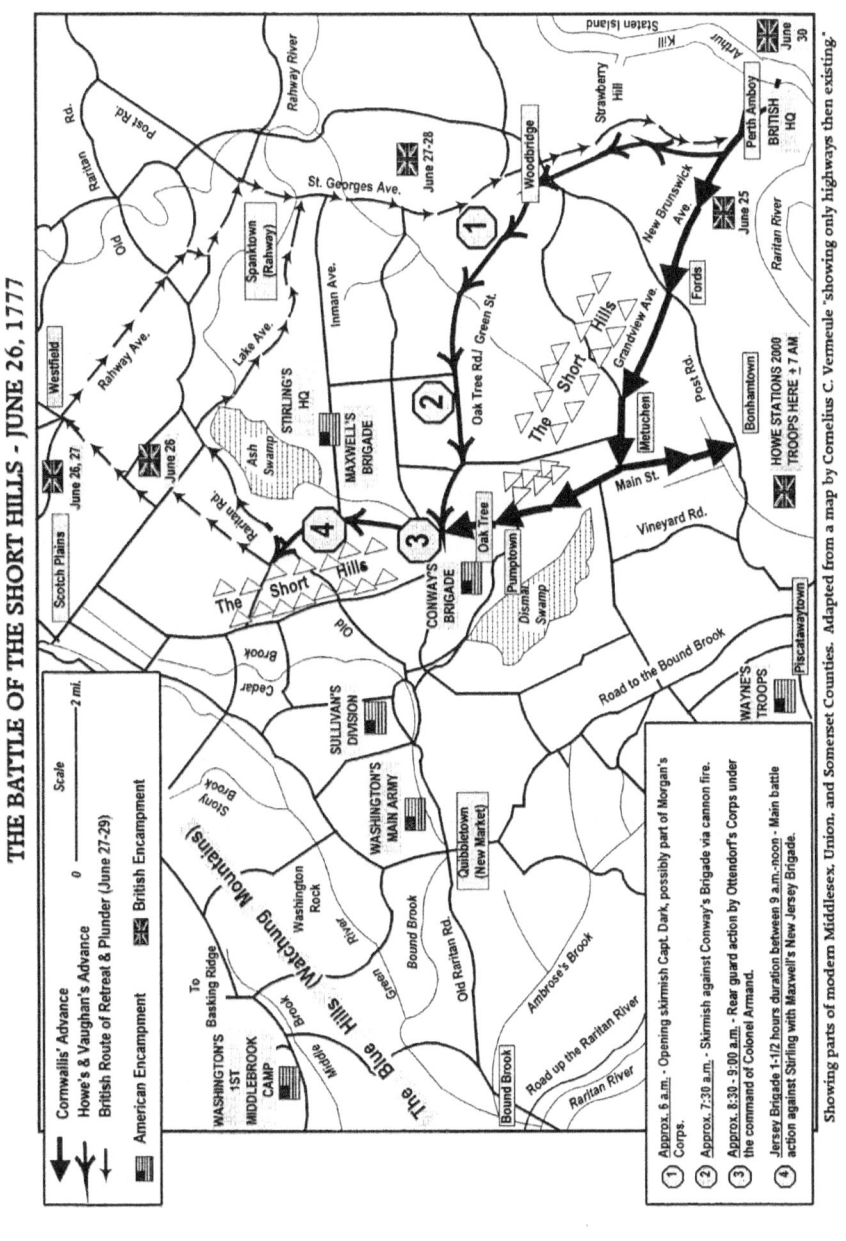

Battle of the Short Hills, June 26, 1777. Parts of present-day Middlesex, Union and Somerset Counties. (*Adapted in 2018 by George Stillman and Ann Walker from an 1885 map by Cornelius C. Vermeule. Roads were modified to show the battle route.*)

Watchung Hills to the flat ground around the Millstone River west of New Brunswick or onto the plains had not succeeded. When the British Army reached Perth Amboy, they began crossing over to Staten Island across the narrow Arthur Kill on barges and on a hastily-erected pontoon bridge.

But then the British commander was astonished to receive a one last opportunity to achieve his master plan. An American deserter alerted them to the fact that Washington had finally come down from the hills, and it now appeared that the American army could conceivably be engaged on the flatlands. This was the opportunity that Howe had been eagerly awaiting for months. He immediately reversed course and began ferrying his troops back from Staten Island. His immediate goal was to commence a surprise onslaught against the Americans, backed by his superior numbers and artillery.

Howe divided his army into two columns, both of which marched on parallel paths toward an area called the Short Hills, low hills west of Rahway lying twelve miles away. His strategy was to form a two-pronged pincer assault to first wipe out Major General Stirling, who was defending the left flank in Scotch Plains. If that move was successful, Washington's main army could then be cut off and prevented from escaping back into the hills at Middlebrook.

It appeared that Washington had moved his army far enough forward that a quick push through Plainfield and Green Brook could envelope the American forces and prevent their escape into the Watchung Mountains. The British could also enter the Stony Brook pass at North Plainfield and move down the Washington Valley through Watchung and Warren, behind the first ridge, to attack the Continental Army from the rear.

Stirling's Brigade of about 2,000 men, vastly outnumbered and outgunned, protected the vulnerable American northern flank and the passes through the first ridge of the Watchung Mountains. They formed a line of defense that stretched two miles—from Edison to the first ridge at Scotch Plains. Stirling stood alone between Washington's unsuspecting Continental Army and the Crown forces, which were over twice their number.

From May 14, when he arrived at Wayne's camp, until the time of the battle that later occurred, General Washington followed the maneuvers of both armies by shifting his observation post from Middlebrook to the iconic Washington Rock in Green Brook. He reached there by riding five miles along the crest of the ridge. Parts of this trail are still visible today, although most of the route has been covered by private homes. Local tradition is that the general spent five days at Washington Rock observing the distant troop movements from this outcropping above the Vermeule Camp. From there it is conjectured that he may have sent orders to the exposed General Stirling using semaphore flags.

The fierce hand-to-hand combat that ensued as the armies made contact started in Edison and spread to Plainfield and Scotch Plains on June 26, 1777. It has come down in history to be known as the Battle of Short Hills. This label causes confusion among modern historians and genealogists, since many assume that the combat occurred in another more modern New Jersey town, Short Hills, a town near Millburn in adjoining Essex County, ten miles away.

Few people have heard of this critical event of the American Revolution, and others have just considered it a minor skirmish barely worthy of mention in the historical record. Many people even drive over the heavily trafficked streets every day during their commute in the area, and even unknowingly hit golf balls from the site of Patriot defense positions on the grounds of the Plainfield Country Club. They are completely unaware of the amazing display of patriotic courage and the intense struggle that raged here on that fateful day. Surprisingly, many details of this significant event are found in the little-known accounts kept by British and Hessian officers, rather than in American sources, although the British side also gave the event little publicity since it failed to accomplish its main objective.

The British Assault Begins

After spending the night consulting with his officers at the Drake House on Front Street in Plainfield early in the morning on June 26, a few hours before the fighting began, Washington climbed to the iconic rock above the first Watchung ridge to observe the action in the lowlands below. The land between the lookout and the Short Hills, a distance of about eight miles, was mostly covered by plowed fields, thus providing an unobstructed view. The fighting at the Short Hills covered a ten-mile area and encompassed the towns of Plainfield, Scotch Plains, Edison, Woodbridge, Westfield and Rahway.

June 26 was a clear but intensely hot and humid day. The weary British and Hessian troops had been marching for a week after leaving New Brunswick to head for Perth Amboy. They had been continuously harassed by snipers concealed in the foliage along their route. They carried heavy backpacks, muskets with bayonets and ammunition, as well as items needed to sustain them: a haversack or large knapsack for food, a canteen of water and a pack or bedroll for spare clothing. The combined weight of this baggage was about fifty pounds. As a result, the Redcoats and their Hessian comrades were exhausted and thirsty.

Sterling's Continentals and local militiamen were personally enraged. The British had been plundering their farms and villages all winter during the Forage War and now they were burning the homes and barns of their neighbors as they passed through the countryside. Thousands of these citizen soldiers had rushed from their homes to harass the British columns as they withdrew from New Brunswick to Perth Amboy, and now they joined up with Stirling's defenders at the Short Hills.

On June 26 Howe's forces were at full strength, a combination of British regulars and Hessian soldiers (mercenaries from the German state of Hesse-Cassel) numbering about 18,000 men. As mentioned, Howe sent his force out from Perth Amboy in two columns, starting out at 1:00 a.m. The right column was led by

Lieutenant General Charles Cornwallis, second in command of the Crown forces in America. The infamous spy, Major John Andre, and the most vilified British Commander of the war, Banastre Tarleton, who would later allow the slaughter of defenseless prisoners after the Battle of Waxhaw in 1780, also led this unit. A second column was commanded by Major General John Vaughan; Howe himself marched with this unit.

The British assault had moved only three miles from Perth Amboy before losing the advantage of surprise at sunrise. As Cornwallis' men moved through Woodbridge, they alarmed a detachment of 150 American troops from Morgan's Rifle Corps led by Captain William Darke. This elite corps of skilled riflemen that had been hand-picked from Virginia regiments was heading toward Perth Amboy on an advanced reconnaissance patrol.

This engagement occurred at about 6:00 a.m. near Edison, at a place called Strawberry Hill (now at the junction of Green Street and Route 1 in Woodbridge). The first American casualty of the day was an African-American boy who had gone to get water for the American officers. Although wounded in the arm, the boy managed to help sound an alert. Runners on both sides quickly reported the location of this initial clash.

Sounds of the First Skirmish Alert a Surprised Gen. Washington

Darke's smaller patrol was opposed by British Captain Patrick Ferguson's company of 250 men, all of whom were equipped with innovative repeating rifles of Ferguson's own design. These guns could be loaded at the breech and had the ability to fire up to six shots per minute.[1] Darke's patrol held off the advance of Cornwallis for about one-half hour, despite being outnumbered and while facing superior weaponry. They continued to hit Ferguson's flanks as the Redcoats broke through and moved up Green Street toward Oak Tree Road in Edison.

The sounds of gunfire from the skirmish at Strawberry Hill alerted the main body of Stirling's defenders. Stirling was then able to warn the unsuspecting Continental Army at Samptown and Quibbletown of an impending assault. At about 7:00 a.m. a scout reported to General Washington that the invaders were only two and a half miles away.

As Cornwallis and Howe's columns advanced down Oak Tree Road, Washington ordered alarm guns to be fired. He also commanded the divisions of Greene, Lincoln, Stephen, and Sullivan to assume defensive positions in the hills at Middlebrook. To the Commander-in-Chief watching the action from his mountain lookout, it appeared that the enemy intended to break through the passes at the first ridge. He immediately ordered alarm guns to be fired, and sent as many men as were available, including the wounded and sick from field hospitals, to set up defense lines at the gaps. General Parsons was positioned at the Bloody Gap (Bonnie Burn Road) at Scotch Plains and General Lincoln at Quibbletown Gap (Warrenville Road) and Stony Brook Pass (North Plainfield and Watchung); Brigadiers Stephens and Woodford held Steel Gap (Chimney Rock Road). The prompt alarm bought the Americans enough time for a fast but orderly withdrawal back to the secure high ground at Middlebrook.

Lacking both arms and enough fighting men, a brash American counterattack at this time could have ended in the large pivotal battle that Howe had been craving. Relying on Stirling's troops at Short Hills to stall the entire British advance, the Continental Army continued its hasty but orderly withdrawal back to high ground. Washington's army had barely survived.

As Darke's pickets fell back along Green Street, Stirling sent out Brigadier General Thomas Conway's Brigade to support them. These defenders were also joined by 700 men from three companies of Pennsylvania-German volunteers led by Major Nicholas van Ottendorf, all under the command of French Colonel Armand-Tuffin. Cornwallis' lead column encountered the first heavy

The Battle of the Short Hills

resistance from this combined force at about 8:30 a.m. at the intersections of Oak Tree Road and Plainfield Road and New Dover Road in Edison. The Americans were supported by three brand-new French brass cannons. This was the first known use of French arms in the war, although a formal alliance had not yet been signed with France.

The British took heavy losses from the barrages fired by the three cannons. After vicious hand-to-hand combat, the vastly outnumbered Patriot defenders were forced back up Oak Tree Road to Woodland Avenue. There they withdrew up the rising ground of the Short Hills through Martin's Woods toward the crest of the hill, now part of the Plainfield Country Club golf course. The Crown forces were able to regroup, and soon charged up the hill in a massive bayonet assault. The valiant Pennsylvania troops took heavy losses—thirty-two men out of eighty in Ottendorff's Corps fell—but they still slowed the British assault. The survivors saved the French cannons and rejoined Stirling's troops on the higher ground.

The rest of the Cornwallis column continued to advance along Oak Tree Road. Their objective was to close the two prongs of the pincer to encircle Stirling by linking up with the left column of 12,000 British troops led by Generals Howe and Vaughan. This column had left Perth Amboy at 3:00 a.m. and headed up New Brunswick Avenue to the town of Metuchen. From there they continued over Plainfield Road in Edison in an effort to approach the Short Hills from the south. This left wing had difficulty mucking through the Dismal Swamp at Samptown (South Plainfield) and did not reach the Short Hills in time to take part in the thick of the fighting or to close the pincer. They finally managed to reach Plainfield Avenue in Plainfield where they joined the rear of Cornwallis force.

Lord Stirling Makes a Stand

Stirling's command post was on the rising ground in Scotch Plains near Ash Swamp, at the junction of Inman and Old Raritan Roads. The Short Hills Tavern and the land of Richard and John Whitehead were also near this site. Stirling's force of 1,798 men, which included General Maxwell's four New Jersey Regiments, formed a defensive line along what is today Tingley Lane and Rahway Avenue in North Edison. At first the Continental troops had a brief advantage. They were able to direct cannon fire over the 800 yards of the rising ground of the Short Hills. As the British advanced along Tingley Lane they took heavy fire from the American artillery concealed behind trees and foliage of the higher ground, now a portion of the Plainfield Country Club golf course.

Colonel Von Minnigerode led one of his four elite Hessian grenadier units on a frontal assault. They bombarded the Stirling's positions on Inman Avenue with a three-pounder cannon. They then circled north around Ash Swamp to block the Americans who had begun to retreat toward the town of Westfield. The Hessians were stopped by "Pennsylvania Dutch" German-speaking troops, who repulsed them with deadly barrages of grapeshot. A single volley of canister (small bullets packed in cases that fit the bore of a cannon) killed six of the advancing Hessians. The Hessian troops were soon reinforced by the Queen's Rangers, who had with them a couple of three-pounder cannons.

The 1st British Grenadiers now joined the assault, supported by at least fifteen cannons. The fighting was extremely intense, with hand-to-hand skirmishing around the four French cannons which repulsed enemy bayonet charges with fusillades of grapeshot. The Grenadiers bravely attempted a bayonet charge up the hill but were driven back. With their superior numbers the British soon gained control of both flanks of the wooded slope rising up to the grounds of the present-day country club. Cornwallis intensified cannon fire and repositioned his infantry for a final assault.

Captain John Finch, a courageous but reckless British officer, charged up to a Rebel cannon alone, and with his pistols he forced the crew to abandon their position. Finch noticed Lord Stirling nearby and shouted, "Come here you damned rebel and I will do for you." In response, Stirling directed four riflemen to aim at the arrogant Redcoat. He was killed seconds later.

Some eyewitness accounts reported that Lord Stirling's horse was shot out from under him. According to local historian George Stillwell, it was actually Colonel Elias Dayton's horse that fell. Dayton, a senior officer in the New Jersey Brigade, had to assume command of the 3rd Pennsylvania Brigade when General Conway acknowledged that he was unfamiliar with the territory. Dayton's advance helped save the rear guard of the 3rd Pennsylvania Brigade and allowed the Jersey Brigade to fall back to another position. These tactics delayed the British for nearly two hours.

The French cannons were lost, then retaken, and then three of them were lost again. French Colonel Armand was able to save one gun as his defenders withdrew. Armand later received a certificate of commendation from Washington for his bravery "at the battle of short hills in Jersey, where 32 of his 80 men were killed." General Maxwell was almost captured by Hessian Grenadiers during the fray.

The American Defenders Withdraw

Taking advantage of their heavy artillery and overwhelming strength in numbers, the British began to surround Stirling's defenders. After two hours the Americans fired a final cannon blast into the encroaching British lines and began falling back into the surrounding woods. The retreating Americans were pursued along Raritan Road, across Rahway Road, and into Ash Swamp in Scotch Plains. As the fighting continued, the famished Redcoats, led by Lord Cornwallis himself, fought their way along Raritan Road toward the concealed American positions in the swamp.

As the Redcoat columns advanced, they approached the house of Gershom Frazee, a carpenter and joiner who was also a staunch Patriot.

His wife was baking bread for the embattled American soldiers as they fell back along the road to Westfield. The mouth-watering aroma of fresh-baked bread wafted through the Redcoat ranks. General Howe approached the modest farmhouse and was met by Frazee's sixty-one-year-old wife Elizabeth. Cornwallis demanded her newly baked bread, and, as legend has it, "Aunt" Betty Frazee replied, "Sir, I give you this bread in fear and not in love." Cornwallis, admiring her audacity, said to his troops, "Not a man of my command shall touch a single loaf." The ravenous soldiers did, however, pillage livestock and household goods before resuming their march. The Frazee house still stands near the corner of Terrill and Raritan Roads in Scotch Plains and is presently being restored and renovated, work sponsored by the Fanwood-Scotch Plains Rotary Club. Aunt Betty died in 1792; she rests with Gershom in the old cemetery at the Presbyterian Church Burial Grounds.

As they withdrew, the thirsty Redcoats guzzled three barrels of applejack that they discovered at Lambert's Mills on Old Raritan Road. James Lambert's later claim for damages lists the amount stolen as twenty gallons.[2] British troops also drank the well dry at the Terry house at Rahway and Cooper Roads on that scorching day in late June. Claims of looting and damage by the enemy forces were later submitted to the state by most residents along the way to Westfield.

The Americans made a final stand at Ash Swamp, and continued to withdraw toward Westfield. By late afternoon the British attack had faltered, and the fighting had broken off. Low on water, and exhausted from the extreme heat and humidity, the British also began marching toward Westfield. From his vantage point on the rock in Greenbrook, Washington could see Stirling's troops. Outnumbered by six to one, they were being beaten back through Ash Swamp and driven toward Westfield. Wagons loaded with American wounded turned west and headed for the pass at Bonnie Burn Road to escape from the enemy.

The victorious British were now in a position to break through one of the passes and attack Washington from the rear. Howe gazed up at the daunting rise of the first ridge of the Watchung Mountains that towered above him. He had to make a critical decision, a choice that could potentially change the direction of the Revolutionary War. Should he risk sending his vast army, trained to fight on open fields, through the narrow pass into the mountainous wilderness? He was tempted. It was a chance to enter the Washington Valley behind the first ridge and then move down Valley Road and Mountain Boulevard in Watchung and Warren to attack the frail Continental Army from the rear at Martinsville. From there, the Crown forces could go on to conquer all of western New Jersey. If he withdrew to New York, he would lose this one-time opportunity.

Cornwallis decided not to risk his army in the heavily defended passes, so he continued a retreat toward Westfield. Washington must have been relieved to see the Crown Forces withdraw, rather than continue to press their assault. In 1871, Reverend Abraham Messler of the First Church of Raritan offered this vivid account of this decisive moment:

The left [column], under Cornwallis, encountered Lord Sterling, and after a severe skirmish, drove him from his position and pursued him over the hills as far as Westfield, where they halted. But the pass in the mountain west of Plainfield [Stony Brook Pass in North Plainfield and Watchung] being guarded, and Washington, like an eagle, perched again upon his eyry, and Sterling beyond the reach of Cornwallis, the British commander saw that the object in view of which his whole manuever had been made, was beyond his reach, turned his face again towards the seaboard; and on the 30th of June crossed over to Staten Island with his whole army. His course was a clear acknowledgment that he was beaten; and that too, by a force far inferior to his own. Both his designs were defeated. He had neither gained an open road to Philadelphia, nor brought on a general engagement; and after maneuvering a month and more, was obliged to change the whole object of the campaign; or seek to gain its end by a circuitous route, in which there was both danger and uncertainty.[3]

Driven from Ash Swamp, the battered Americans turned toward the hills and retreated through Scotch Plains to the gap in the Watchung Mountains at Bonnie Burn Road. From there they returned to the Middlebrook Encampment, through Watchung and Warren, behind the protection of the first Watchung ridge. Howe's aide, Munchausen, counted thirty-seven rebel wagons filled with American wounded struggling up the winding hill on Bonnie Burn Road. For many years afterward the road was known as Bloody Gap.

British Colonel Charles Stuart provided this concise description of the engagement at the Short Hills two weeks after the engagement.

> *On our arrival at Amboy every preparation was made for embarking the Troops, and one Brigade of Hessians had actually embarked. This induced Washington to quit his hold in order to attack us, on hearing which Gen. Howe ordered those Troops to disembark, and at daybreak the next morning march'd in two columns; the right had one, commanded by Cornwallis and Grant, were to pass by Woodbridge to Westfield, and the left by Bowen Town [Bonhamtown] to the same place. Washington's army was drawn up about 3 miles from the Mountains, his left at Sparkston [Samptown], and his right extending towards Boundbrook. "Upon the alarms of our movements, Washington retired to the post he before occupied. Lord Cornwallis, falling in with Stirling, near Matonaking [Metuchen], after a slight skirmish obliged him to retire. In this confusion, we took 60 men, and 3 pieces of cannon; our loss was Capt. Finch, of the Guards, killed, and 30 men killed and wounded, besides 20 men who dropp'd down dead from the heat or fatigue.*[4]

Battle Description by George W. Stillman, Sr.

A local historian, George W. Stillman, Sr., was the most knowledgeable person that I have come across during my extensive research of the Battle of the Short Hills. The majority of his sources are original military documents from both sides, and they provide many previously unknown details. He was my guide in 2010 on a tour of the entire area of the action and graciously shared his comprehensive understanding of the event. He narrated this account as we visited the towns that were encompassed by the engagement and provided an unusually accurate description of the confusing maneuvers of both sides during the action. His account, given to the author on July 7, 2010 is the only existing account of the engagement that describes the role of each unit of both the American and Crown Forces:

I have spent decades researching this forgotten battle. Fortunately, the battle helped save Washington and the American Army. General George Washington had posted the majority of his army around Quibbletown. It was thought that the British Army was planning on moving from New Jersey to Staten Island. Washington hoped there might be an opportunity to strike the British rear guard. Howe, however, was planning a surprise attack against Washington's main forces at Quibbletown. If Washington and the army had been captured, as was the intent of Lord Howe and the British, it likely would have been catastrophic for the new nation regarding its survival.

I believe the New Jersey troops deserve a lot of praise for its brave defense. They only amounted to around 1,000 men. If you include the Pennsylvania Brigade and the early morning advanced troops, Lord Stirling, at most probably had only 2,500. Both the New Jersey Brigade, under General William Maxwell, and the 3rd Pennsylvania Brigade of General Conway, formed Lord Stirling's Division. British commander General William Howe formed his forces into two columns. Lord Cornwallis commanded the right column, which advanced through Woodbridge. Cornwallis' column included the elite Guards Brigade, Hessian Grenadiers, Hessian Jaeger Corps, and a battalion each of the 1st British Light Infantry and 1st British Grenadiers. Two infantry brigades, including Highlanders, were also attached to Cornwallis' column, along with one regiment of British Light Dragoons. The Cornwallis column consisted of approximately 5,000 men. Vaughan's main force, which marched towards Metuchen Meeting House, was joined by Lord Howe with around 12,000 men.

The Oak Ridge Park, which is east of the Short Hills, adjoins Ash Swamp and is likely where General Maxwell's Jersey Brigade was encamped prior to the battle. Lord Stirling had his headquarters at the home of Lt. Edgar of the 2nd Light Dragoons, off Inman Avenue in Edison. Maxwell's men were quartered in nearby farmhouses. General Conway's 3rd Pennsylvania Brigade was closer to Metuchen.

About two miles northwest of Strawberry Hill, in Woodbridge, an American advance guard was surprised by the British light troops in the Cornwallis column. The first American casualty of the day was a young Afro-American boy who had gone to get water for the American officers. Although wounded in the arm, the boy managed to help alert the guard. The Americans in this action consisted of a body of Virginia and some New Jersey troops. This occurred about 5:50 in the morning. Among the men to take part in the resultant retreat was Col. Morgan's newly formed Rifle Corps, which was nearly surrounded, but would manage to recover and later gain fame at Saratoga.

Unfortunately, Lord Stirling received conflicting reports about the exact location of the enemy force and the hard-pressed retreating Americans. To add to Lord Stirling's problem, General Conway reported that he was not familiar with the terrain, so General Maxwell of the New Jersey Brigade was placed in charge of the 3rd Pennsylvania Brigade and was sent to the front to support the retreating American troops coming up Green Street from Woodbridge.

Maxwell halted the Pennsylvanians on high ground, where it forced the advancing British forces into a short skirmish with *musket fire and artillery (possibly from the West Jersey Artillery Company). This action took place on the high ground near Wood Avenue and Oak Tree Road. Upon seeing the large force of Crown troops advancing, Maxwell began a retreat to reach the Short Hills where he hoped to join up with Lord Stirling and the New Jersey Brigade.*

When Maxwell found that Lord Stirling had not yet reached the area of Oak Tree, he decided to continue on towards Quibbletown, leaving a small rearguard detachment from the Pennsylvanians. In the meantime, the British sent out a detachment of the British Light Dragoons to find out where the Pennsylvania Brigade had gone. The rearguard of the Pennsylvania detachment was temporarily surrounded and captured at Oak Tree.

But suddenly the Jersey Brigade under the command of Col. Elias Dayton, with the 1st and 3rd New Jersey Regiments, advanced around the edge of the Short Hills and surprised the British. The first artillery fire came from the East Jersey Artillery Company, which startled the British. Musket fire from the advancing Jersey line forced the British Light Dragoons to release their prisoners and retreat. Much of the following serious fighting took place between Oak Tree Road, Woodland Avenue, Tingley Lane, Inman Avenue, and Old Raritan Road. This area had been designated as the Short Hills Battleground Historic District.

The 2nd and 4th New Jersey Regiments were on the left of the line and were accompanied by the East Jersey Artillery Company and Capt. Jones' Independent Pennsylvania Artillery Company. Col. Dayton and the Jersey line also discovered five companies of British Light Infantry hidden in the woods and forced them to retire deeper into the wood line in the Short Hills. The Pennsylvania Brigade, however, was split off from the New Jersey troops once the Pennsylvanians reached Old Raritan Road and moved west toward what is today South Plainfield. That left the New Jersey units to confront the growing strength of Cornwallis' advancing units.

The Hessian Minnegerode Grenadiers, having reached Oak

Tree, were ordered to advance to their right flank with other Hessian Grenadier Battalions to the rear in support. Cornwallis posted the British Grenadiers and Royal Artillery on high ground to the left of the Hessians. An American corps known as Ottendorf's Corps under command of Col. Armand was hit hard by the returning embarrassed British Light Dragoons. Armand was able to save an American cannon but lost 32 casualties out of Armand's 80 men.

With the increasing sound of battle growing louder by the minute, the smaller body of the Jersey Brigade fell back to hilly terrain towards the current Plainfield Country Club. The fight near Oak Tree had lasted about thirty minutes. Fighting now raged into the hills to the north and east near the Laing Farm on Woodland Avenue to the Short Hills Tavern on the corner of Inman Avenue and further to the east towards Tingley Lane, Inman Avenue and Rahway Road.

Lord Stirling had positioned the New Jersey Brigade with the 1st and 3rd New Jersey Regiments on the right, near where the Hessian Jaegers, possibly joined by the British Light Infantry, attempted to get around the Jersey Line. Lord Stirling also positioned the 4th New Jersey Regiment in the center of the Jersey Line. with two French artillery guns from the East Jersey Artillery Co. and the 2nd New Jersey Regiment on the left with two French guns of Jones' Artillery Co. as they awaited an attack from the Hessian Grenadiers and British Foot Guards. The 4th New Jersey Regiment and East Jersey Artillery guns concentrated on the Hessians while the 2nd New Jersey Regiment and Jones' guns fired on the Guards.

Lord Stirling stubbornly stayed with the guns at the center of the New Jersey line. A young officer of the Guards named Capt. Finch rode off in advance of the Guards, forced away the artillerists at the center guns and threatened Lord Stirling. Lord Stirling directed the fire by some marksmen who mortally wounded Capt. Finch. The Americans recaptured their guns, but Lord Stirling soon realized he needed to retreat to prevent the New Jersey

Brigade from being surrounded. He ordered the New Jersey troops to begin a controlled retreat while he used the American cannons to buy time for their defense. Overwhelmed by superior numbers, the patriots began falling back along the Old Raritan Road into Scotch Plains.

Three cannons were lost in the intense hand-to-hand combat but the phase of the battle under Lord Stirling had lasted a precious two hours. The severe fighting slowed along the Old Raritan Road near the Frazee House in Scotch Plains.

Howe's other column, under General Vaughan, eventually arrived at the battle to join on the rear of Cornwallis, but the joined forces were no longer necessary since the main body of the American Army had fallen back into the Watchung hills between Middlebrook and the Scotch Plains gap.

The Patriots retreated toward Westfield and then onto the Watchung heights while trying to block the Crown forces to allow Washington more time to reach the safety of Middlebrook. They then moved up Bloody Gap in Scotch Plains, now Bonnie Burn Road, to continue behind the first ridge through Watchung and Warren to Middlebrook.

The British and Hessians admitted to seventy men killed, wounded, or suffocated by the heat, at the Short Hills and thirteen taken prisoner by the Americans. Perhaps the most reliable number of American losses was published later in the Continental Journal a record of the daily proceedings of the Continental Congress. It reported losses of three field pieces, twenty killed and forty wounded. Accurate casualty numbers, however, were difficult to obtain due to differences in reports. For example, Capt. Ewald of the Jaegers estimated Crown force losses as 130 while some reports claimed American losses were around 200.

Following the battle, Howe's Army retired to Staten Island on June 30, 1777, where he had begun the New Jersey campaign one year before. Howe never personally returned to New Jersey but began his fruitless planning, known as the Pennsylvania

Campaign, to capture Philadelphia. And just a few days after the battle, the patriots, had survived to celebrate the first anniversary of the young nation's independence on July 4 1777.

While a tactical defeat for the Americans, the Battle of The Short Hills was really one of the most significant strategic victories of the Revolutionary War. The resistance by Stirling's outnumbered force at Short Hills and Ash Swamp saved Washington's army. His valiant stand against a vastly superior force allowed time to move the Continental Army back to the safety of the hills at Middlebrook and to avoid a British victory that would have doomed the struggle for independence. This engagement and the strong opposition of the New Jersey Militia once again caused the British to fail in their attempt to gain entry into the interior of New Jersey. Their final departure was a welcome relief to the long-suffering citizens of the state and a tribute to the courage of the defenders.

The Battle of Short Hills was the last effort to engage Washington's forces early in the summer of 1777. Both armies then moved south for the Philadelphia Campaign. The main British Army would not return to New Jersey for the remaining six years of the war. The courage of Stirling's brigade, the citizen soldiers of New Jersey, and the natural fortress provided by the Watchung Hills prevented the downfall of our new nation. Washington's strategy at the Battle of Short Hills earned him the respect of his adversary Lord Cornwallis. Four years later, after his surrender at the final decisive American victory of the war at Yorktown, the British commander said, "But after all, your Excellency's achievements in New Jersey were such that nothing could surpass them."

Chapter Eight

The Rampaging Retreat

The Redcoats Leave New Jersey

HIS MAJESTY'S VICTORIOUS troops emerged from the Battle of Short Hills in the late afternoon of June 26. They were exhausted from the sweltering day which had begun that morning at 1:00 a.m. They were also irate. Even the lowest ranked private knew that the opportunity to draw General Washington into a major engagement had been squandered, and that the New Jersey Campaign had failed. Their rage boiled over as they continued northeast, now unopposed, from Scotch Plains to Westfield, a town less than five miles away. The weary and aggravated troops ruthlessly looted and destroyed the farms along the way. Frustrated British officers began allowing their normally disciplined and restrained men to devastate civilian property. Claims filed by many residents detail the theft and extensive damage that ensued.[1]

This extract from the *Pennsylvania Evening Post*, a Philadelphia newspaper, describes the depredations committed as Howe's army moved along Broad Street to Westfield: "Morristown, July 5, 1777, The British army burnt, stripped and destroyed all as they went along. Women and children were left without food to eat or raiment to put on. Three hundred barrels of flour were sent down towards Westfield and the Ash Swamp, by order of his Excellency [General Washington] to be distributed among the poor sufferers. The enemy destroyed all the Bibles and books of divinity they came across."[2]

The old village of Westfield had been beyond the reach of British foraging parties during the winter months of 1777 and was spared the destruction that occurred at the Amboys, New Brunswick,

Bound Brook, and on the farms along the Raritan River. But now this unblemished town, with its abundance of produce from the rich surrounding farmland, was ripe for predation.

The Westfield Meeting House, a Presbyterian Church built in 1728 at East Broad and Mountain Avenues, frantically tolled its bell to warn of the approach of the thousands of ravenous Redcoats and Hessians. Residents heeded this warning to hastily evacuate, and hurriedly fled to safety in the hills of present-day Mountainside. Unfortunately, there was little to no time to gather possessions or livestock.

Despite marching fifteen miles from Amboy and fighting at the Short Hills in the extreme heat, Howe's energized troops continued their merciless pillaging in Westfield.[3] They camped on the grounds of the Presbyterian Church, which they regarded as being offensive to their Church of England and as a symbol of radical patriotism. Jacob Ludlow, a New Jersey militiaman, reported, "The British filled the church with sheep and put a ram's head in the pulpit and slaughtered a great number of hogs, sheep and cattle. They threw down the bell from the steeple and slaughtered sheep and cattle in the building."[4] The picturesque present-day church on the site dates back to 1862. It is the same height as the older structure, the same one in which the British army bivouacked on the night of June 26, 1777.

Other jubilant Redcoats camped overnight in Westfield along Willow Grove Road and Central Avenue. It was a long night of feasting and drinking. The reveling soldiers cooked freshly slaughtered livestock in camp kettles hung from the branches of walnut trees along East Broad Street.[5] Many of the Redcoats spent the night in the homes and shops of the Patriot residents.

Based on claims later filed by Westfield homeowners with the State of New Jersey, Crown troops looted and destroyed at least ninety-two houses in Westfield and plundered a staggering 11,000 individual items from the homes.[6] They also tore down over 2,000 fence rails and posts for fuel in order to cook their food. In addition, they stole all the cash in the town. Curiously, a slave found the opportunity to escape from his American master during the melee. Pitifully, he is

The Rampaging Retreat

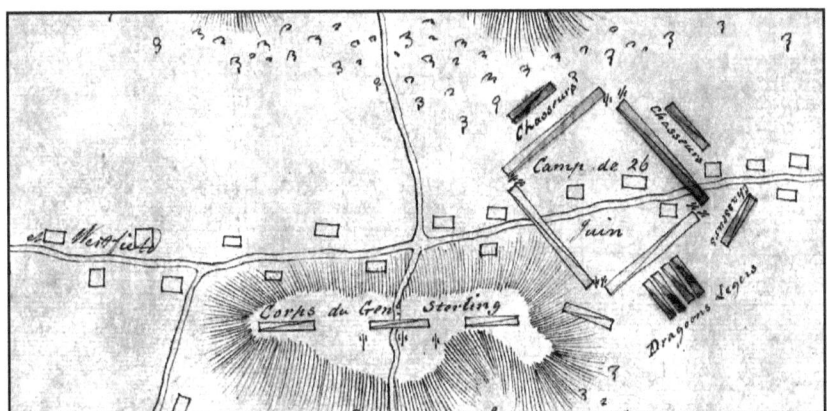

Detail of the British encampment at Westfield. From "Plan de l'affaire de Westfield & du camp de Raway" by Friedrich Adam Julius Von Wangenheim, 1777. (*Library of Congress, Geography and Map Division*)

listed fourth on an inventory of plundered items, after "2 excellent milk cows, 1 two year old heifer, and 2 large fat hogs." Although he was valued at 80 pounds, his name and age are not recorded.[7] On June 27, even British Major John Andre lamented that "The spirit of depravation was too present on these marches."[8]

The British army abandoned Westfield by 9:00 a.m. the next morning to march south along Grand and Rahway Avenues. They camped that night along the Rahway River. Along the way they absconded with about 500 cattle and several wagonloads of looted property. Major Andre confirmed this in his journal: "At 9 in the morning we marched by the left bringing with us about 60 prisoners picked up in different places and driving the cattle we met on the road"[9] The prisoners were American soldiers captured at Short Hills, as well as some unfortunate civilians apprehended along the route to Rahway. All were eventually imprisoned in the notorious Sugar House Prison in New York City, where many soon died from starvation. Others were transferred to prison ships moored on the East River, off Brooklyn. During the war, 11,500 prisoners died there due to overcrowding, contaminated water, starvation, and disease.

When American forces reentered Westfield two days after the

Crown forces departed, they were appalled at the rapaciousness of the invaders. Colonel Israel Shreve of the 2nd New Jersey Regiment filed this report:

> They made shocking havoc, destroying almost Everything before them, the house where Gen. How stayed which was Capt. Clarks he promised Protection to If Mrs. Clark would use him well and Cook for him & his Attendance, which she Did as Chearful as she Could, Just before they went off Mr. How [sic] Rode out, when a No. of his soldiers Come in And plunder[ed] the Woman of Everything in the house, Breaking And Destroying what they Could not take Away, they Even tore up the floor of the house, this Proves him the Scoundrel, and not the Gentleman, Gen. Lesley took his Quarters At Parson Woodruffs [and] Protected his property in Doors, the Doctor fled [but] his Wife and famaly Remained, the meetinghouse a Desent Building they made a sheep pen of threw Down the Bell, and took it off[f]..., they Drove of[f] All the Horses, Cattle, Sheep & hogs they Could Git, – I saw many famalys who Declared they had Not one mouthful to Eat, [nor any] bed or beding Left, or [a] Stitch of Wearing Apparel to put on, only what they happened to have on, and would not afoard Crying Children a mouthfull of Bread Or Water Dureing their stay.[10]

Alexander Hamilton, while serving as an aide to General Washington, wrote this summary of the battle and its aftermath to William Livingston, providing further evidence that after the engagement at Short Hills the victorious Redcoat forces appeared to completely lose all sense of honor and chivalry during their retreat:

> Head Quarters Camp, at Middlebrook, June 28th 1777. Lord Stirling's party was near being surrounded: but after a smart skirmish with the enemy's main body, made their retreat good to Westfield, and ascended the pass of the mountains back of Scotch Plains. The other parties after the skirmish on their flanks came

off to join the main body and take possession of the heights. The enemy continued their march towards our left as far as Westfield, and there halted. In the meantime, it was judged prudent to return with the army to the mountains, lest it should be their intention to get into them and force us to fight them on their own terms.

They remained at Westfield till the next day, and perceiving their views disappointed have again returned to Amboy, plundering and burning as usual. We had parties, hanging about them in their return; but they were so much on their guard no favourable opportunity could be found of giving them any material annoyance. Their loss we cannot ascertain; and our own, in men, is inconsiderable, though we have as yet received no returns of the missing. I have no doubt they have lost more men than we; but unfortunately, I won't say from what cause, they got three field pieces from us, which will give them room for vapouring, [boasting] and embellish their excursion, in the eyes of those, who make every trifle a matter of importance....[11]

From Rahway the Redcoats marched seven miles back to Perth Amboy, continuing to sack, burn, and loot Patriot homes along the way. On June 30, 1777, the Crown forces once again crossed back over the Arthur Kill to Staten Island. Their campaign in New Jersey has lasted eight months. Two days after the British left, General Washington moved the army from Middlebrook to an even more secure place—Pompton—sixty miles further inland. This First Middlebrook Encampment, punctuated by the Battle of Short Hills, had lasted thirty-five days, from May 28 to July 2, 1777. The Continental Army would return seventeen months later, in November 1778, for its winter encampment.

In late July 1777, the American Army marched south again after learning that a British invasion fleet was on its way to Chesapeake Bay, and that Philadelphia would need to be defended. The defense of New Jersey was once again left to its valiant local militia.

The British Campaign in New Jersey during the first half of 1777 was not only a military failure. Ravaging the countryside to sustain their army alienated a large formerly-neutral population, and destroyed the loyalty of many Loyalists who were indiscriminately punished. While engaging in their futile prolonged forage war the British suffered significant losses and desertions without any major engagement, while Washington's army remained largely intact.

The disheartened Parliament in London expected a major victory and the destruction of the American Army in New Jersey in 1777. The alternative strategy, to abandon New Jersey and to capture Philadelphia with an amphibious invasion from the head of Chesapeake Bay, did not provide the Crown forces with the decisive blow they needed to end the war. While the forthcoming capture of Philadelphia had a psychological impact on the Patriots, it did not embolden any Loyalists or determine the ultimate outcome of the conflict.

These developments stoked criticism of Howe's leadership in Parliament. Other commanders in the British Army also began to question his competence. Munchausen, Howe's loyal aide, wrote, "Because our recent expedition seems to have failed, several rumors are spreading. Almost everyone blames General Howe, but I am convinced that many would have done worse, but none better."[12]

In October 1777, Howe sent his letter of resignation to London. He complained that he had been inadequately supported in that years' campaigns. His resignation was accepted, and he sailed for England in May 1778. General Henry Clinton took over as commander-in-chief of British armies in America, and left Philadelphia for an overland march to New York. On the way his army was confronted by Washington's army at the Battle of Monmouth near modern-day Freehold, New Jersey. The Crown forces were driven from the field of battle and evacuated to New York City from Sandy Hook.

The inability of Howe to force a decisive battle or outmaneuver General Washington in New Jersey during the First Middlebrook Encampment caused a three-month delay for Crown forces. This was due to the long sea voyage to the northern tip of Chesapeake Bay where the Philadelphia Campaign was launched. This maneuver by the British Army away from New York City and the Hudson River Valley contributed to the eventual halt and defeat of John Burgoyne's army at Saratoga since Howe's forces were in no position to support that advance.

Remarkably, General Washington correctly recognized that while tradition held that the main objective of an army was to go on the offensive and aggressively attack the enemy, his goal was simply to survive. The British failure to lure Washington down from the stronghold in the Watchung hills would disturb them for the remaining six years of the war. The evasive strategy of the Continental Army won the Foraging War in the winter and spring of 1777, and the avoidance of a final decisive battle was critical to the army's survival. The strategy of remaining on the high ground of the Watchung Mountains, while the British controlled the Raritan River Valley and the access to New York Harbor and the Atlantic Ocean, proved to be a decisive success. Washington's army was ready to fight the British again after their engagement at Short Hills.

The Scene Today

The Short Hills ascend the slope along the southern edge of the Plainfield Country Club. Stirling's defenders occupied the high ground where Patriot cannon fired down the slope toward Oak Tree Road. Today, this site still offers a commanding view over the former battlefield and the surrounding landscape. Descriptions by early settlers in the 17th century describe wooded lots and cultivated fields. The battlefield area was farmland until the early 20th century, when suburban development spread into the area. The northern portion of the land was converted to a golf course in the first quarter of the 20th century and this location is now occupied by the country club.[13]

The view from the country club sweeps over built-up suburbia, busy roads and a few tree-covered areas, and in no way resembles the appearance of the 18th century agricultural landscape consisting of a patchwork of plowed fields. The panoramic view in the opposite direction, looking northwest to the first ridge of the Watchung Mountains, also provides a outlook over a heavily developed area with most evidence of man-made activity obscured by trees.[14]

Historical descriptions and maps show the major roads over which the troops on both sides advanced and retreated. Today they still exist as busy urban streets which still follow their 18th century routes. Oak Tree Road, the thoroughfare that the Cornwallis column followed west from Woodbridge, and where much of the fighting occurred, does not appear on any early maps. It was likely little more than a dirt path. It's route can only be traced by damage claims filed by property owners, and modern geographic mapping.[15] Bonnie Burn Road, known after the battle as "Bloody Gap," passed by a house built in 1768 that was, sadly, recently demolished. It serves as a heavily-trafficked conduit between Route 22 in Scotch Plains and Interstate 78 in Watchung. Present day Main Street in Metuchen and Plainfield Road in South Plainfield mark the route of advance for the southern or left column of the Crown forces, which arrived too late to participate in the fighting.

PART II

The Second Middlebrook Encampment, November 1778 - July 1779

Chapter Nine

Why Middlebrook?

THE CONTINENTAL ARMY left Valley Forge in the spring of 1778. In June it clashed with the Crown Forces on a sultry hot day at the Battle of Monmouth in Freehold, New Jersey, proof that Washington's forces could stand up against the British Army, considered at the time to be the best in the world. After that battle the Americans headed north to the Hudson Highlands, and the British proceeded back to New York City. From the Highlands Washington's army could block the Hudson River to prevent a northern incursion by the British Army.

As the winter of 1778 approached, Washington decided to move the Continental Army back to Middlebrook, his former secure encampment. For the first time since the war began in the spring of 1775 the Commander-in-Chief was optimistic. An alliance had been signed with France, the Patriots had held the field at Monmouth, and they had then reoccupied Philadelphia. There was cautious optimism and hope that the war could end favorably for the American cause.

After the Battle of Monmouth, the Crown forces withdrew to Sandy Hook and sailed over to New York City. Washington marched north to the Raritan River and rested his troops in the New Brunswick area, the former stronghold of the British. There they celebrated the Fourth of July with a grand parade and a fireworks display. The thundering noise made by the fireworks terrified the British on Sandy Hook. They believed that they were being attacked, and realized that if this were the case they were now trapped on that narrow spit of land. Eleven days later the American army continued north twenty miles to Scotch Plains and Newark. On July 15, Patriot forces reached Haverstraw, New York in the Hudson Highlands, forty miles up the river from the city.

Later that month Washington moved the army across the Hudson River to White Plains. With the Continental Army at full strength in Westchester County he could try to keep the British penned up in New York City. The enemy remained on the defensive there, and in fact had not strategically extended its occupation zone in the city over the past three years.

Washington decided to reorganize the command of the American army in September of 1778. General William Alexander, Lord Stirling, was given command of the Virginia Regiments. General Israel Putnam was given the Massachusetts Line. Generals Horatio Gates, Anthony Wayne, Benjamin Lincoln, Alexander McDougall and Baron Johanne DeKalb were all were placed in command of their own divisions.

During the last six months of 1778 there had been no further movement by the British in New York. Certain that they would remain in the city, the Continental Army prepared to move to its own secure winter encampment in the hills of central New Jersey. Washington announced his intention to leave Westchester County to his General Staff on October 14:

> It will readily occur to you that the following particulars are to be considered, the security of the army itself, its subsistence and accommodation, the protection of the Country, the support of our important posts, the relation which ought to be preserved with the French Fleet if should it should remain where it is.
>
> I have been waiting impatiently for the movements of the enemy to come to an issue that might ascertain their intentions for the winter, which had hitherto prevented my taking the present step; but the uncertainty in which they still continue involved, and the advanced season of the year, will no longer admit a delay in fixing on a plan of the general disposition of the army in winter quarters.[1]

By the end of November a final decision had been made to return to the Middlebrook Encampment in what is now the Bridgewater,

New Jersey area. From this high ground on the first ridge of the Watchung Mountains the American forces could quickly mobilize and move anywhere to protect New Jersey or assume an aggressive stance and threaten the British on Staten Island and in New York City.

In the 18th century most military activity in the western world came to a stop during the winter months. It is difficult to understand why operations would be suspended for an entire season regardless of the army's location, strategic position or level of activity. However, winter encampments were necessary due to a combination of harsher weather at that time, human energy requirements and lack of protective clothing and gear that could not function well in low temperatures. Therefore, a season to rest, recuperate and repair equipment was essential. Traditionally, hostilities would come to a standstill in the late fall and both sides would go into permanent winter camp for several months. A secure and easily defended place was selected, cabins were built, and a supply chain established. Many officers and men would go on furlough. Combat resumed in the spring. General Washington explained the importance of winter camps:

> *The principal objects of the Commander in chief for collecting the Army together in a Cantonment of repose for the winter, were to keep alive the spirit of emulation amongst the different corps, to confirming the habits of Discipline which had been acquired last Campaign; and improving the internal police of each to the highest degree of perfection. Although the severity of the season may not for some time admit of the troops being drawn out for exercise and maneuvering yet the greatest attention might, and should be in the meantime paid to the establishment of perfect regularity, economy, and good order, throughout the line of the whole army.*[2]

Washington learned the value of this seasonal strategy early in the war. In December 1775, half of the forces of the fledgling

American army were annihilated while trying to storm the bastion of Quebec City during a raging snowstorm. A tragic 200-mile retreat followed that disaster. The entire Canadian campaign proved to be a weather-related calamity for the Patriots. They never forgot this dreadful episode.

There were exceptions to the traditional routine pause during the winter months, the most famous of which occurred when Washington crossed the freezing Delaware River on Christmas night to surprise the Hessians at Trenton, an action followed up a week later with two victories at Princeton. But when the fighting broke off a week later, in early January 1777, both sides headed directly to winter camps in New Jersey. The triumphant Patriots marched to Morristown where they camped until the end of May 1777. The king's army withdrew to New Brunswick and remained there until late June.

The most notable American long-term winter encampments took place at Morristown, Pompton and Middlebrook in New Jersey, New Windsor in New York and Valley Forge, Pennsylvania. The British forces had a comfortable and festive winter in the captured rebel capital of Philadelphia in 1777-1778 season. Most of their main army spent the remaining five years of the war secure in New York City. Historians persistently depict battles and eulogize victories of the Continental Army, but often overlook the fact that troops were encamped most of the time. The extensive seven-month-long winter encampment at Middlebrook, neglected by history, is an obvious example.

Why Middlebrook?

Middlebrook, a natural fortress, was the site of two major encampments of Washington's Continental Army during the Revolutionary War. These occurred in the early summer of 1777, and during the winter of 1778-1779. The encampment was located in present day Bridgewater Township and Martinsville, New Jersey and covered an area of five miles. Portions extended to today's Somerset County towns of Warren Township, Somerville. Bound Brook, Raritan, South Bound Brook, Manville, Millstone and Franklin Township.

The campground was centered northwest of Bound Brook, where Vossellor Avenue and Route 22 intersect today, and along the gap in the first ridge of the Watchung Mountains, now Chimney Rock Road. It included a large area in front of and behind the first ridge of the Watchung Mountains from Chimney Rock Road to King George Road and up to Dock Watch Hollow Road in Warren Township. A separate station at Pluckemin, seven miles west, served as a base for the Artillery Brigade.

The encampment derived its name from a small 18th century town named Middle Brook, situated where that stream enters the Raritan River. The Middle Brook flows through the Washington Valley into Martinsville behind the first ridge. It then runs through the gorge at Chimney Rock and joins the Raritan River at Bound Brook. The site of the hamlet was where Thompson Avenue and Route 28 meet today, near Somerset Patriot Baseball Park, and where Interstate 287 crosses the Raritan River. Over the years, the area was soon encompassed by the town of Bound Brook, so it cannot be found on any modern maps.

The heights of Middlebrook on the first Watchung ridge looked down on the Raritan River Valley. The fertile valley, settled with only a few small villages, extended southeast over eight miles of rich farmland to the town of New Brunswick. The area was sparsely settled, with only a scattering of farms. It would change into a no man's land during the Forage War in the first turbulent months of 1777. Bound Brook, at the junction of the Raritan River and Middle Brook, was a tiny river port that served the agricultural hub of the region.

George Washington officially selected Middlebrook as an appropriate winter camp in October 1778.[3] He knew the place well. He had marched through the pass at Chimney Rock Road earlier in 1777 on his way to Morristown after the Battle of Princeton.

Later that year the Continental Army had camped there during the First Middlebrook Encampment, which lasted from May to July of that year.

Middlebrook had obvious logistical, topographical and geographical military advantages. The high ground of the Watchung Mountains, which rose sharply hundreds of feet above the plains of the Raritan Valley, made it a natural fortress that could be easily defended. In an age when all cannons and supplies were dragged by men and horses this elevation made it impregnable. If overwhelmed, the American Army could live to fight another day by hastily withdrawing deeper into the safety of the hills and forests of Morris County.

The site had a thirty-mile panoramic view of the no man's land that stretched over the plains of central New Jersey. This vast area extended east to Raritan Bay, northeast to Elizabethtown and southeast to the Amboys and New Brunswick. In 1777 the British maintained a force of as many as 18,000 troops only eight miles away in New Brunswick. When they ventured out into the countryside their movements could be detected from observation posts on the heights above Bound Brook, and from other rocky perches that ran north along the first Watchung ridge.

Middlebrook was also at the junction of a road network that provided access to outposts and supply depots. Roads led north to Morristown, south toward Princeton and Trenton, or north and southeast toward Elizabethtown or New Brunswick. Defenders from the camp could easily rush to protect adjacent mountain passes at Chimney Rock, Greenbrook and Watchung. The place was near enough to New York City so that be activities at the British Headquarters could be monitored by visual observation. Also, the many local residents who acted as spies in the city could easily communicate with Washington. The area was supported by the strong and active New Jersey Militia regiments from other counties. In fact, these citizen-soldiers made up most of the fighting force during the Forage Wars of early 1777.

The entire population of Somerset County was only about 10,000

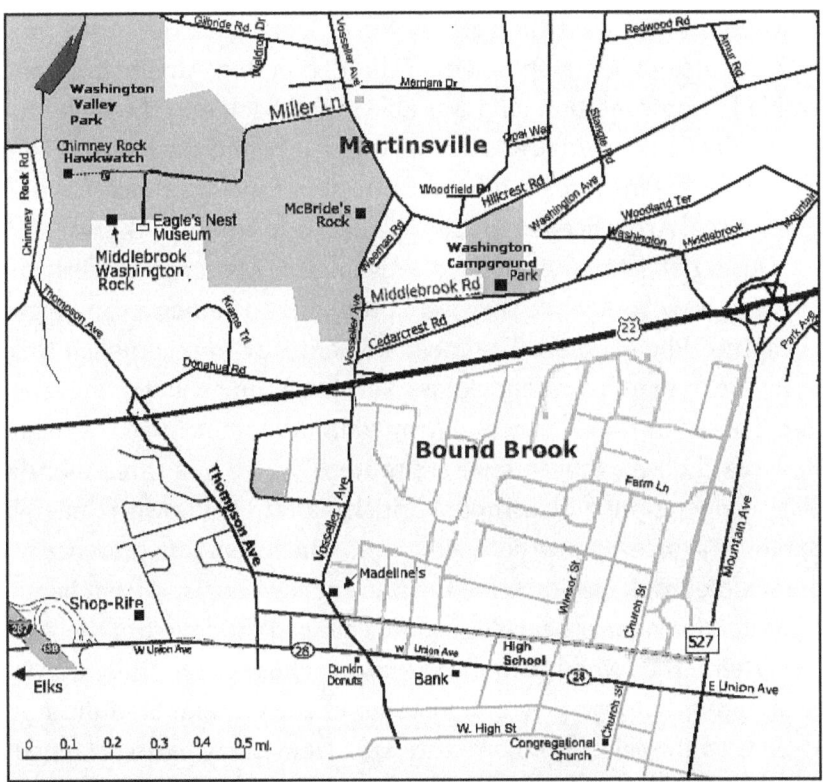

Detail of the Middlebrook Encampment showing rock lookout sites, overlaid by today's landmarks and roads. (*Courtesy of Don McBride, Martinsville, NJ*)

residents at the time. The farmers of the surrounding Washington and Raritan Valleys was generally patriotic and receptive to providing supplies and labor to the army. While it could not entirely support the large contingent of the Continental army for the seven months of the winter encampment, the agriculturally rich country could contribute significantly to ease the insufficient food supply. The area also had a plentiful water supply and abundant trees that could be used in the construction of huts and for fuel.

The British master plan during the first half of 1777 was to provoke the American Army to move down from the high ground to the open plains and to engage in a decisive traditional European style battle fought with linear tactics. Confronting a vastly superior foe would be a certain defeat for the Patriots forces which were less than one half the size of their opponents. The debilitated American Army avoided a confrontation, and was able to fall back into the hills and protect its flank when it was almost trapped before the Battle of the Short Hills in June 1777. Middlebrook provided the sanctuary that the Continental Army needed to remain intact and continue the war.

During the summer of 1777 Washington observed the British forces in New York while they were preparing to launch a campaign to capture Philadelphia. The most efficient way to accomplish this would be to march overland across New Jersey, but concern over the threat posed by the Continental Army at Middlebrook forced General Howe to choose a longer, safer sea route. This led to a three-month delay in operations and disrupted British plans to merge with General Burgoyne's forces in northern New York. The forbidding presence of the Middlebrook bastion served to prevent New England from being separated from the other colonies and changed the course of the war.

When the Middlebrook site was chosen for the winter encampment of 1778-1779 one member of the General Staff did not agree with the selection. Lord Stirling, a New Jersey native, claimed that the food supplies and forage in the surrounding area could not possibly sustain the entire army. He recommended that the American forces be spread out. A detachment could be assigned to Essex County, and the remainder divided up throughout various New Jersey towns. Although this suggestion had merit, it was rejected by General Washington. He was more willing to tolerate supply problems in order to maintain discipline and keep his main army intact.[4]

About 300 acres of the Middlebrook Encampment have been preserved as a National Historic Site. Much of the heavily wooded tract remains unchanged from the days of the war. The Washington Camp Ground Association owns twenty-five acres in the heart of the area. The five mile radius of the encampment encompassed the colonial bridges across the Raritan River in Bound Brook. Today, there are several carefully restored houses in the area that were occupied by Washington and his generals.

At present, parts of the encampment lie under Routes 22 and Interstate 287. One of the few historic signs commemorating the place now stands in a residential area, at the corner of Chimney Rock and Gilbride Roads in Bridgewater. Much of the remaining undeveloped land occupied during the first encampment in 1777 is owned by the Somerset County Park System. Future residential building on the surrounding land is very likely.

Chapter Ten

The American Army Arrives

The Winter Encampment of 1778-1779
From White Plains to Middlebrook

IN JUNE 1778, after the encouraging Battle of Monmouth, the Continental Army moved north to spend the summer in White Plains, New York. At this point, General Washington's army had the greatest strength that it would ever have at any time during the entire war. The American force totaled close to 17,000 troops. In September it was reorganized into six divisions, or brigades, which were commanded by major generals Putnam, Gates, Stirling, Lincoln, De Kalb and McDougall.

On October 26, 1778 Commander-in-Chief Washington directed General Stirling to issue a general resolution from the Continental Congress. The decree would eventually govern conduct of the army after it reached the next winter encampment at Middlebrook. With few exceptions this virtuous declaration was obeyed for the entire seven-month bivouac of the army at Middlebrook during the winter of 1778-1779.

> Whereas, religion and good morals are the only solid foundation of public liberty and happiness; Resolved, That it be, and hereby is earnestly recommended to the several States to take the most effectual measures for the encouragement thereof, and for the suppression of theatrical entertainments, horse racing, gaming and such other diversions as are productive of idleness,

dissipation, and general depravity of principles and manners.

Resolved, That all officers in the army of the United States, be hereby strictly enjoined to see that the good and wholesome rules provided for the discountenance of profaneness and vice, and the preservation of morals among the soldiers, are duly and punctually observed.

In consequence whereof, the Commander-in-Chief of the army in this Stare, directs, that strict obedience to the foregoing resolves be paid by all officers and soldiers within the same Francis Barber, Adj.-Gen.

Washington reaffirmed these orders after the encampment was occupied,

The Commander-in-Chief approves of the order issued by Major Gen. Lord Sterling during his command at the camp, and thanks him for the endeavor to preserve order and discipline, and the property of the farmers in the vicinity of the camp. He doubts not but the officers of every rank, from a just sense of the importance of securing to others the blessings they themselves are contending for, will use their utmost vigilance to maintain those privileges and prevent abuses, as nothing can redound more to their personal honor and the reputation of their respective corps. General orders, Alexander Scammil. Adj.-Gen.

In October, General Washington also decided to divide the army into two parts for the duration of the 1778-1779 winter encampment. Most of the troops would be assigned to Middlebrook. From there they could deploy instantly to protect all parts of New Jersey or block the Hudson River by reinforcing the Hudson Highlands. British-occupied Staten Island and New York were near enough to be threatened or attacked from Middlebrook. The second part, consisting of the effective New Jersey Brigade, would be stationed separately, near Elizabethtown, to defend against assaults from New

York. Other troops, stationed at Danbury, Connecticut could defend the Hudson Highlands or repel an invasion from Manhattan. These camps and their outposts created a forty-mile defensive arc around the British-held New York City.

In November, brigade commanders assigned to Middlebrook were ordered to leave White Plains and undertake the sixty-mile trek to the winter camp in New Jersey. On November 20, Baron De Kalb left with the Maryland Brigades and several Tory prisoners, and Brigadier General Muhlenberg did the same. Five days later, orders were given to Brigadier General Anthony Wayne and to Colonel Daniel Morgan to march out. As a result, by late November all major units of the American Army were on the way to Middlebrook. General Washington traveled south as far as Elizabethtown, but in mid-November he abruptly returned to the Hudson Highlands. Intelligence had alerted him to the fact that a British task force had left New York City to strike his vulnerable columns as they crossed the Hudson River at Kings Ferry near Haverstraw.

Meanwhile, an enemy flotilla with fifty-two ships and a large number of troops had been sighted navigating up the Hudson River. The Redcoat plan was to sever the American lines of march by catching the Patriots at Kings Ferry as they crossed the river to Rockland County in New York. The Americans would be defenseless, with their troops on one side of the river awaiting the supplies and artillery being ferried across from the other side. Other units, now strung out for miles in the New Jersey countryside, would be easy prey. The Crown forces then planned to continue to advance north to capture the American stronghold of West Point.

Fortunately for the Patriots, the fleet was delayed by conflicting orders. The British decided not to follow through with this plan and their armada returned to New York. If their strategy had succeeded it would have been a crushing defeat for the Americans and might have ended the war.

During the first week of December in 1778, after they were certain that the enemy had withdrawn, all American regiments resumed the march to Middlebrook. On November 27 General Washington wrote that "the whole army is now in motion to the place of winter quarters, if no unexpected interruption happens, the whole will be over this 30th instant."

General Washington finally arrived at the Middlebrook camp on December 11, 1778. His arrival is documented in a December 13 letter which he sent to the President of Congress: "...I did not reach this place till late on the 11th, since which I have been much employed in attending to the dispositions [plans] for hutting the army.[1] At this point in time an American line of defense was in place for ninety miles. It extended from Middlebrook through Elizabethtown in New Jersey, through the Hudson Highlands to West Point and east to Danbury, Connecticut.

The Army Arrives

Troops arrived at Middlebrook at the end of November and into early December, 1778. The regiments dispersed north of the village of Middle Brook and into the gap in the hills at Chimney Rock Road. Other settled along the base of the first Watchung Ridge where Route 22 passes today. The camp covered the greater part of Somerset County, New Jersey and encompassed the Raritan and the Washington Valleys. Its geographic center today is a about at the junction of Routes 22 and Interstate 287.

About 10,000 troops reached the encampment by mid-December. A combination of enlistments, furloughs and illness caused the total number during the encampment to fluctuate. In January 1779, 8,875 men were numbered on the muster rolls. However, by deducting those who were either sick, on furlough, or assigned outside the camp as picket guards, only 4,738 remained prepared for active duty. By the end of March the total number had increased to 7,029. With the arrival of new recruits by the end of April, the total again increased, this time to 9,804.

The sparse civilian population of the area could not independently support the thousands of soldiers now stationed in their vicinity. Each state had the responsibility for feeding, clothing, and equipping its own troops. State governments were the main sources of supply for the regiments they sponsored. They varied in their commitment to the Continental Army, and often gave preference to their own militias. Actually, the states simply did not yet have the necessary support functions, infrastructure, or experience to adequately supply the army on such a large scale at one location. Therefore, supplies reaching Middlebrook were often unreliable and insufficient. Provisioning problems arose as soon as the troops arrived, and plagued the army for the entire duration of the encampment.

Location of Regimental Campgrounds

Lessons learned from the previous dreadful winter at Valley Forge influenced the placement of the regimental camp sites at Middlebrook. With the inactive British army securely settled in New York City, there appeared to be no immediate danger of a major enemy advance along the route from New York to Philadelphia. So Washington believed that the Raritan Valley would be a safe location for campgrounds. Less accessible places occupied during the 1777 encampment in the mountains, or in the more remote Washington Valley behind the first ridge, were avoided.

After assessing the ease of defense and their available wood and water sources, each division independently selected a campsite within the perimeter of the campground. The main army, consisting of the Virginia, Maryland, and Pennsylvania Brigades and the Delaware Regiment, settled in the Raritan Valley between the base of the Watchung Mountains and the Raritan River. The adjacent mountains in the rear provided a buffer that sheltered the area from severe weather coming from the north and there was an abundant supply of lumber for construction and firewood on the densely wooded land. The Virginia and Maryland camps on the west side of the Raritan River were within easy reach of roads leading toward New York and the New Jersey coast.

The Virginia troops consisted of three regiments commanded by Generals Charles Scott, William Woodford, and Peter Muhlenberg. They arrived at Middlebrook on December 9 and were posted just west of the gap where Middle Brook Creek flowed out of the mountains along what is now Chimney Rock Road. The Virginia position extended along today's Foothill and Steele's Gap Roads in Bridgewater Township.

Steeles Tavern was a well-known landmark at the time of the encampment. It provides us with a reference point and enough evidence to accurately place locations of the regiments. Orderly books often show warnings forbidding the soldiers from visiting the public house. At the time it was managed by Adrian Manley and was located on the road to Pluckemin, today's Route 202, about a mile from the junction of Old York Road in Bridgewater. The place is often referred to as Steele's Gap on old maps

In 2018, local residents in the neighborhood raised $50,000 toward preserving the historic place where 2,000 Virginia soldiers were stationed. The 36-acre site is located at the intersection of Foothill Road, Twin Oaks Road and Steele's Gap Road, and includes the Wemple Homestead.

The land was sold to developer Steven Lang, who proposed building homes on the land. David Stempien, a fifty-year Bridgewater resident, lives on Steele's Gap Road next to the Wemple Homestead. He joined with his neighbors—Bob Vaucher, Fred Shandor, and Brendan Burns—to convince Steven Lang that the property had significant historical value. Lang recognized the importance of the site and allowed the land to be preserved. The Wemple property is the wooded area to the north and east of the elegant farmhouse that can be seen from Foothill Road. Local tradition claims that Steeles Tavern was located adjacent to a spring there that served as a water source for the troops.

The Continental Army arriving at the Middlebrook Encampment, December 1778. (*U.S. National Archives*)

This place lies on the Washington-Rochambeau National Historic Trail, the route taken in 1781 by General Rochambeau's French Army on its way to Yorktown, Virginia. This historic trail runs from Rhode Island to Virginia, and passes along Washington Valley Road, Adamsville Road, Steele's Gap Road and Foothill Road.

Dr. Robert Selig, project historian to the National Park Service for the Washington-Rochambeau Revolutionary Route National Historic Trail Project, believes there are potentially "thousands of artifacts" to be unearthed on the Wemple property. The land has been transferred to Bridgewater Township and The Crossroads of the American Revolution, National Heritage Area. Their mission is to maintain a place where the public can experience New Jersey's

Revolutionary past and see where troops actually marched, foraged and prepared for battle. Crossroads is working with the township and local historians to install interpretive signage and a network of trails to mark the site.

The Maryland Brigade commanded by General William Smallwood and Colonel Josias Hall, also arrived on December 9. It was posted east of the Chimney Rock gap and extended past Vosseler Avenue to Mountain Avenue in Bridgewater, along what is now Route 22. Numerous references in regimental orderly books about duties assigned to this unit indicate that the Maryland Line was located south of the Virginia camp closer to the village of Bound Brook.

The Pennsylvania Brigade, commanded by Generals Anthony Wayne, Arthur St. Clair and William Irvine, was the last to arrive in the late December. It was assigned by General Greene to a site across the Raritan River from the main camp, a more remote location separate from the other troops. This was the only area remaining that could support a large campsite. The Pennsylvania camp extended from along Main Street in today's Manville and followed Millstone River Road into the town of Millstone. It covered present day Whalen and Okin Streets and ended at Frech Street near Van Nest's mill.[2] The Corps of Artillery, led by General Henry Knox, encamped just north of the crossroads village of Pluckemin, seven miles west of the Virginia and Maryland main camps.

Historian Benson Lossing visited the Middlebrook encampment in the early 1850s. Based on information provided by the older inhabitants of the area he reported that the campground was "on the gentle slope from the plain to the steep acclivities of the mountain in the rear of Middlebrook [Bound Brook]."[3] Other evidence of the camp location was shown in a map drawn by General DeKalb. He placed the main encampment along the base of the first ridge behind Bound Brook, with some huts positioned on the slope of the hill, while others reached down to the level land below.[4] Present day Route 22 bisects this area.

When General Greene first explored the Middlebrook area as a possible winter camp in November 1778, he wrote to Washington," I shall immediately go out to take a view of the ground toward Steeles Tavern to fix upon a place for hutting." As early as March of 1778, Washington ordered the construction of a signal beacon at Steele's Gap. Verification of the location of the Virginia Line comes from the proximity of a signal tower erected there. Beacons were usually erected in high places close to where the troops were camped that built them.

Washington's Cartographer Verifies Campground Sites

Robert Erskine, General Washington's cartographer, drew several maps of the Second Middlebrook Encampment. The drawings show the supply network of local roads and structures, illustrate elevation, and provide evidence of the locations of regimental camp sites. His locations are supported by the plentiful evidence found in orderly books, officer's letters, and other military papers. Erskine clearly shows the positions of the three divisions. His map traces a road between Middlebrook village and Bound Brook that ran north from Old York Road. It is marked "to the Middebrook Valley past the Maryland Camp."[5] Using today's landmarks, these maps also place the Virginia Camp near Chimney Rock, southeast of the site of Steeles Tavern on the sloping terrain of the first ridge.

Erskine shows the Maryland Division located to the south of the Virginians in an area between Chimney Rock and the town of Bound Brook. A series of trenches remained visible there until Route 22 was widened in recent years. Several smaller defensive fortifications guarded the front and flanks of the encampment but have disappeared over time. A large redoubt that protected the rear and flanks of the cantonment still exists in Martinsville, in the Washington Valley. It can be visited on West Circle Drive, off Spring Run Road, a mile west of Martinsville center.

General Washington Called to Philadelphia

On December 22, 1779, less than two weeks after his arrival in Middlebrook, General Washington was summoned to Philadelphia to report to Congress. He been with the army continuously for the previous thirty months and had not been home to Mount Vernon for three-and-a-half years. In Philadelphia, he soon grew restless and impatient with the inertia of the legislature and was annoyed that members of Congress lived lavishly while the army was enduring hardships at Middlebrook. He stayed in the city for the next forty-five days, during which he complained to Benjamin Harrison, Speaker of the Virginia House of Burgesses:

> *If I were called upon to draw a picture of the times, and men that I have seen, heard and in part known, I would say that idleness, dissipation and extravagance seem to have laid fast hold of most of them. Speculation, peculation [misuse of funds entrusted to you] and insatiable thirst for riches seem to have gotten the better of every other consideration and almost every order of men. Party disputes and personal quarrels are the great business of the day whilst the momentous concerns of an empire, a great accumulating debt, ruined finances, depreciated money and lack of credit, which in their consequences is the want of everything, are but secondary considerations and postponed from day to day, from week to week, as if our affairs bore the most promising aspect.*[6]

The general was joined by his wife Martha in Philadelphia, and they were the guests of Henry Laurens, the President of Congress. While in town, they attended parties and Washington had his portrait painted by Charles Willson Peale. Washington declared that this period was the first relief he had experienced since joining the army in 1775. George and Martha returned to Middlebrook on February 8, 1779 and moved into the Wallace House in Somerville.

The unrewarding trip to Philadelphia had depressed the

general, and he was soon confronted with a sea of problems back at the encampment. He found that discipline at Middlebrook had deteriorated and desertions had surged. Droves of unruly men used any excuse to avoid work. Officers were taking long unauthorized furloughs. Frequent court martials with harsh punishments were not reducing infractions. But, worst of all, the amount of supplies arriving was not sufficient to sustain the army.

Chapter Eleven

Settling In

Hutting, the First Priority

THE ORGANIZED APPROACH to hutting at Middlebrook was a major advance over the shoddy construction that had taken place at Valley Forge the previous winter. Over 2,000 men had perished from exposure related illnesses there. On the other hand, only three deaths from exposure are recorded among the nearly 10,000 troops that stayed at Middlebrook from November 30, 1778 until June 3, 1779. Improvements in shelter and sanitation provided by the strictly enforced hutting plan at the camp made that winter a success in survival for the American Army. Fortunately, during most of the encampment the weather was mild, and the temperatures remained above freezing. Snow fell only in late December and early January. Fortunately, this foul holiday weather was the only major precipitation of that winter.

After five months of idleness in New York the men were well rested. General Washington directed General Nathanael Greene to lay out the locations of huts and to start collecting building materials. On October 29, 1778, just before the army had arrived, he ordered Greene to "collect boards, stone and such material as are requisite to make barracks comfortable."[1] Once the army showed up, their first task was to construct shelter. Hut building began the first week in November before any accumulation of snow. Construction was facilitated with lumber often coming from trees cut down on the building site.

Building materials soon became scarce, as Lord Stirling had predicted, and Greene was disappointed with the progress made in setting up the camp. On December 4, he wrote to Charles Pettit, Assistant Quartermaster General, "On my arrival here I find little in preparation made for the accommodation of the army, but I am told materials are coming in from all quarters, and am in the hope in a few days a great plenty will be collected. Forage and provisions are very scarce, greater will be necessary to get us a seasonable supply."[2]

With the recommendation of General Washington, Congress had appointed Baron von Steuben Inspector General of the army on May 5, 1778. When the baron toured the Middlebrook camp he was appalled at the deplorable condition of huts, weapons, and equipment. There had been no sanitation plans. Men relieved themselves anywhere and when an animal was slaughtered it was stripped of its meat and the remains was left to rot where they lay.

Steuben quickly laid out a plan for straight rows of huts. General Washington began conferring each day with the baron and even personally supervising some of the hut construction. A grid of company and regimental streets was designed, and dwellings were laid out with separate designated areas for officers and enlisted men. Officer's huts and a parade ground were placed in the front row. Kitchens and latrines were placed on opposite sides of the camp that sloped downhill.

To prevent the dampness that had occurred at Valley Forge, dwellings were not sunk into the ground or covered with earth. Roofs were covered with boards or large shingles and each man built a bunk with enough room to store his musket. Ditches were dug on the high side of each hut row to drain the ground, and to keep the huts dry. Paths between the rows and the parade ground were kept clear of any debris.[3] The sanitation standards and camp layout established at Middlebrook would be followed by the army for the next century-and-a-half.

Washington's Life Guard, an elite cadre responsible for his personal protection, were stationed in Raritan near his residence at the Wallace house in Somerville. They were the first unit to complete their huts, and they moved in on December 16. Elijah Fisher, a soldier who served in the Guard, reported in his diary, "we finished out huts and we left our tents and moved into them"[4]

Winter arrived at the camp with a heavy snowstorm on Christmas Day. General Wayne reported "a very severe storm from Christmas to New Year." He had reason to be concerned. Temperatures during that week were near zero. The Pennsylvania Line, delayed by the threat of the British advance up the Hudson River, had arrived last. Until their huts were built they suffered through chilling nights in their tattered tents. Some men, without any type of shelter or extra blankets, were completely exposed to the chilling wind.

Wayne warned Joseph Reed, the representative from his state in Congress, that many officers were requesting furloughs, and he doubted that many would return afterwards. He reported on December 28, "All the Pennsylvania line at this inclement season exposed to wind and weather in their old tents are quite destitute of blankets and without hats...so naked as not to be fit to appear on parade."[5] After arduous days felling trees to build the huts, the men of his brigade were the last units to finish erecting them. At the end of January, General Wayne reported that his Pennsylvania huts were finally nearing completion and that the entire camp was finished. According to Surgeon James Thacher, the Pennsylvania huts were not actually completed until early February.[6]

When the Maryland and Virginia regiments had completed their shelters General Stirling reported that, despite the severe cold, those units were in great health and good spirits.[7] The more substantial barracks at the artillery park in Pluckemin were not ready for occupancy until the early part of February.

During this difficult time civilian anger over billeting soldiers in their houses was avoided. All men and officers, except for a few senior commanders, were forbidden to seek civilian housing. This kept many officers in tents until January. Generals Washington and Wayne considered sending them home for the remainder of the winter, but feared that most of them would not return.

Farmers living near the camp were enraged when the

Pennsylvanians stormed through the countryside tearing down fence rails and outbuildings for firewood. During the worst weather officers began leaving the camp to find more comfortable quarters. Stern warnings were again issued imploring them to remain with their men.[8]

Life in Tent City

Huts at Middlebrook were built using the natural materials found on the wooded hills surrounding the campsite. The only materials provided from army stores were boards for the bunk beds and a few expensive nails. Hand tools were scarce. There was no glass for windows nor iron for hinges. The huts were a standard size, sixteen feet by fourteen feet, with walls seven feet high. These same dimensions are found at other large winter encampments during the war. The shelters had a peaked roof and a fireplace.

The logs used for the thick walls of the huts were notched on the ends and dovetailed together. The spaces between the logs were filled with clay and straw. The roof was a single slope with enough pitch to shed the snow and rain. It was made of timbers and covered with hewn slabs of wood and old straw. There may have been boards for some floors, but most were dirt.

Soldier's huts had one window and one door. Officers' quarters usually had two windows. The windows and doors were formed by sawing out an opening in the logs and mounting the part removed on wooden hinges. Sometimes the window hole was covered with oiled paper to let in light. The door was at one end of the hut. At the opposite end there was a chimney built of small logs. Both the inner and outer sides of the chimney were covered with clay or plaster to prevent the wood from igniting.

The furnishings of the huts were sparse. The only light was provided by candles which were issued to each man with his rations. The tallow for them came from the cattle slaughtered for camp use.

Settling In

The open fireplace was the only source of heat, and poor ventilation sent as much smoke into the room as it did up the chimney.⁹ Replica huts can be seen today at Valley Forge, Jockey Hollow and New Windsor.

As an example, regimental orders regarding hut construction were later issued for the New York Brigade on December 7, 1779:

Lieut. Gray is appointed to superintend the building of the huts with respect to the dimensions thereof to lay out the ground between the soldiers, officers and field officers huts and to see that the tools are properly divided and taken care of, for which purpose he is to take receipts from the officers to whom they are delivered, and if they are borrowed, the officers to whom they are first delivered are to be accountable. The whole regt. is to be employed in building the huts by companies except the guards and one sergt. And 8 men who are to build the Colo. Hutt. The officers in genl. except Mr. Gray will see that the men do their duty the whole is to turn out to work every morning at eight o'clock if the weather will permit until the whole is finished.

Continental Army Surgeon James Thacher, stationed at Middlebrook, described the details of hut construction and layout at the encampment:

February 3 [1779] - Having continued to live under cover of canvas-tents most of the winter, we have suffered extremely from exposure to cold and storms. Our soldiers have been employed six or eight weeks in constructing log huts, which at length are completed, and both officers and soldiers are now under comfortable covering for the remainder of the winter. Log houses are constructed with the trunks of trees cut into various lengths, according to the size intended, and are firmly connected by notches cut at their extremities in the manner of dovetailing. The vacancies between the logs are filled in with plastering consisting of mud and clay.

The roof is formed of similar pieces of timber, and covered with hewn slabs.

The chimney, situated at one end of the house, is made of similar but smaller timber, and both the inner and the outer side are covered with clay plaster, to defend the wood against the fire. The door and windows are formed by sawing away a part of the logs of a proper size, and move on wooden hinges. In this manner have our soldiers, without nails, and almost without tools, except the ax and saw, provided for their officers and for themselves comfortable and convenient quarters, with little or no expense to the public. The huts are arranged in straight lines, forming a regular, uniform, compact village. The officers' huts are situated in front of the line, according to their rank, the kitchens in the rear, and the whole is similar in form to a tent encampment.

The ground for a considerable distance in front of the soldiers' line of huts is cleared of wood, stumps and rubbish, and is every morning swept clean for the purpose of a parade-ground and roll-call for the respective regiments. Line officers' huts are in general divided into two apartments, and are occupied by three or four officers, who compose one mess. Those for the soldiers have but one room, and contain ten or twelve men, with their cabins placed one above another against the walls, and fitted with straw, and one blanket for each man.[10]

All tools were issued by the Quartermaster Department. Jacob Weiss, Deputy Quartermaster, made frequent requests for tools from other army facilities. Weiss' letters, now safely stored in the archives of the Lehigh County Historical Society, were published in 1956 in the society's Proceedings. The requests provide an interesting glimpse into the world of 18th century military procurement in the Continental Army. On December 16, 1778 Weiss requested, "Broad Axes, Adzes Claw or Carpenter's Hammers, 12 or 15 Cross Cut Saws with cross cut and Hand Saw Files and also Saw Setts, 10 or 12 Saddles with the prices or distinguishing whose Merchandize, About 2 Tun

Barr Iron as wrote for including that for Mr. How, 10 or 12 Barrs Steel suitable for new Steeling Axes &c.-And a good Stove with pipes agreeable to dimensions."[11]

Living conditions of the soldiers became more comfortable when they moved into huts. The horrible suffering at Valley Forge the previous winter was not repeated. In March, 1779, General Washington reported to Marquis de Lafayette, "The American Troops are again in Hutts, but in a more agreeable and fertile country, than they were last winter at Valley Forge; and they are better clad and more healthy than they have ever been since the formation of the army."

On April 1, new recruits and soldiers returning from furlough began swarming into the Camp at Middlebrook. In a frantic attempt to find shelter, the new arrivals began building rickety huts on any available space around the camp. The crude dwellings did not conform to specifications. Many interfered with drainage and posed a sanitation hazard. General Smallwood of the Maryland Regiment reluctantly ordered all these huts to be torn down and rebuilt to conform to camp regulations.[12]

In 1779 spring arrived at Middlebrook a couple of weeks earlier than usual. Nathanael Greene, Quarter Master General, had managed to keep the troops reasonably well fed and clothed during the winter months. There were weeks when food was scarce, but the soldiers never starved as they had at Valley Forge or at Morristown a year later. In the early spring a welcome shipment of new uniforms arrived from France.

Chapter Twelve

Everyday Life

Soldiers at the Middlebrook Winter Encampment

THE MOST MOMENTOUS consequence of the seven-month winter encampment at Middlebrook was the maturation of the American Army. It transformed itself from a starved, disorderly, and incompetent rabble into a disciplined fighting force of professional soldiers. We see evidence of national unity replacing loyalty to each state. At Middlebrook, the Continental Army consolidated into a power that could match the larger, better financed, and well-trained British forces. This was all achieved by the effective orgnizational skills of General Washington and his senior commanders. It occurred despite a broken supply chain, suffering from cold and lack of food, expiring enlistments and widespread desertions.

Newly appointed Inspector General Baron von Steuben is the one person that should be credited with turning the volunteers into a impressive army in order to achieve this level of development. At Middlebrook he continued the program of progressive military training that he had begun the previous winter at Valley Forge. Instruction in the essentials of drills, tactics, and discipline were constant.[1] While at Middlebrook Steuben prepared *Regulations for the Order and Discipline of the Troops of the United States* commonly known as the "Blue Book." Its drills and tactics were followed by the United States Army for the next seventy years.

Other significant events occurred at Middlebrook that had a profound effect on the outcome of the war. On February 1778, ten months before the encampment, France formally recognized the United States with the signing of the Treaty of Alliance. Hostilities

Everyday Life

Baron von Steuben drilling troops at winter encampment, a painting titled "Baron von Steuben at Valley Forge," by Augustus G. Heaton, 1907. *(National War College of the United States, National Defense University, Washington, D.C.)*

had soon followed after Britain declared war on France the following month. The strengthening of this alliance occurred at Middlebrook with an official visit of M. Girard, the French Ambassador to America. During the visit he pledged to increase the flow of supplies from his country, and proposed that the French fleet coordinate their operations with the Continental Army. Generous shipments of muskets and clothing from France soon began arriving at the camp. Washington and his staff planned the next major offensive of the war. The campaign against the Iroquois Nations in the north was launched as the encampment concluded in July 1779.

As the worn out soldiers marched into camp after months of being either on the move or in combat, they could finally anticipate a time to rest. Many men would be allowed to return home on furlough. Interesting visitors would arrive at the camp and holidays would be celebrated. Improved shelter, heating and sanitation would prevent another disastrous Valley Forge experience. There would be time to engage in simple recreational activities.

But camp life was also tedious and boring, a nonstop round of drilling, guard duty and "fatigue work." Fatigue work involved such onerous chores as burying the remains of butchered livestock, digging latrines and making cartridges. The monotonous daily routine was summarized by Captain Lewis of the Virginia Line. "For the future the fatigue parties to parade at seven o'clock in the morning and return at eleven to their dinners and parade again at two. Then came supper, evening prayers and tattoo."[2]

When the troops first went into winter quarters they were busy. Huts were being built and roads were being surfaced. Entrenchments and redoubts were constructed around the perimeter of the camp. But when these projects were completed, the days became filled with the repetitive drilling, training, and everyday chores. Firewood was collected, muskets were repaired and cleaned, meals cooked, clothes washed and mended, and huts cleaned. Seasoned veterans best endured the boredom and routine at the grim winter camp. They were hardened into the privation and mind numbing routine of soldiering.

Guard or picket duty required staying awake all night on the perimeter of the camp to watch for an enemy attack or to report any suspicious activities or threats. Falling asleep on guard duty was a serious offense and was punishable by twenty lashes. Guard duties were assigned each day and were expected to be performed in all kinds of weather. Regimental orderly books, the daily record of activities in camp, show that vigilance and strict discipline were constantly maintained, regardless of harsh weather. For the eager young men who had signed up the previous year, the hardship and monotony of life in service did not resemble their fantasies of adventure and military glory.

Routine marching maneuvers were practiced on parade ground, a large area of level hard land in an open field. Musters, inspections, drills and punishments were also regularly held on the parade ground. At times, these events involved the entire army, and were often attended by General Washington and visiting members of the Continental Congress.

The winter encampment at Middlebrook in 1778-1779 began

Everyday Life 137

Winter Encampment. *(Illustration by Benson Lossing from* The Pictorial Field Book of the Revolution*)*

during what may be considered the nadir of America's military and political fortunes. With little funding and a discouraged public, weary of war, officers began to resign, and desertions were rampant. The Continental Army dwindled from 8,875 men in January 1779 to 7,029 by April. Washington warned Congress that the states were not filling their quotas of new recruits. Volunteers became insufficient to fill the dwindling ranks. Pleas for longer enlistment periods became urgent as the war ground relentlessly on. In the spring Middlebrook was stripped of men to prepare for the campaign against the Iroquois Nations.

To encourage reenlistments among veterans in the camp, a cash bounty of $200 and a new uniform were offered. Long furloughs, up to three months, would also be given to those agreeing to stay on. Due to the lack of manpower at that time, all other furloughs regardless of duration were shortened, and finally they were eliminated. In desperation, the Continental Congress reluctantly directed the states to draft men from their militias for one year's service in the Continental Army.

This decree was not well received and was soon circumvented. When more affluent men were drafted they were allowed to arrange for substitutes to serve in their place. Slaves, servants and apprentices were offered as stand-ins for their masters. Draftees also "purchased" petty criminals and the indigents from law officers to take their places. British deserters and prisoners of war also found their way into the Continental Army. This conscription enveloped the young and helpless men from the least advantaged levels of society. Curiously, this had a positive long-term effect. These men, from the lowest socioeconomic level in the nation, men with few other options in life, eventually evolved into an effective army of full time, long term soldiers.

This first draft was poorly enforced and administered. The states continued to rely on land and cash bounties to attract recruits. By March 1779, General Washington was so alarmed by the lack of troops at Middlebrook that all men on furlough were ordered to report back to camp before May 1. By the end of April, however, the tide turned. There were not enough huts to accommodate new recruits and soldiers returning to camp from furlough.[3]

Diversion and Jollification

The soldiers at Middlebrook had little opportunity for fun or recreation until their huts had been completed. At that point they soon discovered a wide variety of diversions to break the monotony of the daily routine. Distractions ranged from playing an innocent game of ball to witnessing a fellow soldier's punishment. References in diaries and orderly books mention the many ways in which they were entertained.

The most prevalent diversion by far was consuming alcoholic beverages. The soldiers played catch with a ball and card games. "Pitch penny" was a popular game. When the men had no money, a button was substituted for the coin. Marksmanship contests were encouraged, as was fishing and hunting. There was ice skating in the winter and swimming during the summer in the Raritan and Millstone Rivers, and in the Middle Brook which coursed through the campground. Taking a bath in a river or lake was a noteworthy occasion but in one case led to a problem:

> *Complaints having been made by the inhabitants situated near the mill pond that some soldiers come there to go into swimming in the open view of the women and that they come out of the water and run to the house naked with a desire to insult and wound the modesty of female decency, 'tis with concern that the general finds himself with the disagreeable necessity of expressing his disappropriation for such beastly conduct.*[4]

The New Jersey Militia and Line regiments stationed at Middlebrook had visits from family and friends living in the area. This was a common practice at all the winter camps. Some of the outings took place with little regard to danger. Mahitable Pride, the wife of Sergeant James Pride of the 5th New York Regiment, was at Fort Montgomery, New York sharing a picnic lunch with another soldier's wife. Suddenly, they heard the sound of gunfire. It was the start of a British assault on the fort.[5]

Many of the restless young men were actually exuberant teenagers. This sobering entry appears in a regimental orderly book of the New York Brigade at Middlebrook in the spring of 1779.

> *The Comdg. Officer is under the disagreeable necessity of informing his Soldiers of that which in his Opinion their own good sense and time of Service ought long before this to have Convinc'd them of the unpropriety of...their noisy unsoldierlike conduct when in their Tents...seems to increase daily to such a degree as to render the situation of their Officers verry disagreeable & expose themselves to the ill natured observations of the Soldiers of Other Regiments in their passage thro' the Camp. He therefore expects that they*

will behave with more propriety for the future, and moderate their Mirth so as to render the situation of those whose duty Obliges them to be near not so disagreeable as it has hitherto been, he has not the least Inclination to lay them under any restrictions that will check their Mirth provided kept within due bounds but on the contrary would rather encourage it as far as it is consistent with good order & Military discipline.

Grim Diversions

Spectacles of corporal punishment provided macabre entertainment during the uneventful days in camp. Attendance at punishments was usually mandatory. It was believed that witnessing hangings and floggings and other brutal penalties would set an example. Firing squads and "running the gauntlet" required audience participation.

The gauntlet required the condemned soldier to be stripped to the waist and to pass between a double row of solders who struck him with sticks or switches. A low-ranking officer walked in front of the guilty party with a sword to prevent him from running. The offender might also be dragged through the camp by a rope around his hands or prodded along by a pursuer.

Frequent last-minute reprieves added theatrics to the punishment drama.[6] When a man was sentenced to death, he was brought bound and blindfolded to the side of a fresh grave. The firing squad lined up and snapped back the locks on their muskets. At the last possible moment, at the height of spectator apprehension, there was an unexpected interruption and a reprieve would be granted. The suspenseful event ended well for the victim, but would serve to haunt the witnesses. The now contrite and terrified offender would likely avoid further offenses.

Ebenezer Wild, a Corporal in the 1st Massachusetts Regiment, described such an execution in his diary:

The culprits marched to the place of execution to the strains of the "Dead March," with their coffins carried before them. The guilty men stood in front of their entire brigade. Their death sentences were read in a loud voice. Their graves were dug, the coffins were laid beside them, and each man was commanded to kneel beside his grave while the executioners received their orders to load, take aim. At this critical moment a messenger rushed in with a reprieve which was read aloud.[7]

Virtuous Middlebrook

Directives issued by Congress in 1776, the first year of the war, established the standards of conduct for the army:

It is earnestly recommended that all officers and soldiers diligently attend divine service and all officers and soldiers who shall behave indecently or irreverently at any place of divine worship shall, if commissioned officers be brought before a court martial there to be publicly and severely reprimanded by the President.[8]

General Washington insisted on piety among the troops, and made this known through his general orders. At Valley Forge, on May 2, 1778, he had issued an order which included this sentence:

To the distinguished character of a Patriot, it should be our highest glory to add, the more distinguished character of a Christian.[9]

Beginning in 1776 a chaplain was assigned to each regiment, and Congress promoted devotion by issuing 20,000 Bibles to the army in September 1777. The Commander-in-Chief, through Lord Sterling, caused the following resolutions of the Continental Congress to be published to the army on October 26. 1777, a month before its arrival at Middlebrook:

> Whereas, religion and good morals are the only solid foundation of public liberty and happiness;
>
> Resolved, That it be, and hereby is earnestly recommended to the several States to take the most effectual measures for the encouragement thereof, and for the suppression of theatrical entertainments, horse racing, gaming and such other diversions as are productive of idleness, dissipation, and general depravity of principles and manners.
>
> Resolved, That all officers in the army of the United States, be hereby strictly enjoined to see that the good and wholesome rules provided for the discountenance of profaneness and vice, and the preservation of morals among the soldiers, are duly and punctually observed.
>
> In consequence whereof, the Commander-in-Chief of the army in this Stare, directs, that strict obedience to the foregoing resolves be paid by all officers and soldiers within the same - Francis Barber, Adj.-Gen.

Washington reaffirmed these orders after the encampment was occupied:

> The Commander-in-Chief approves of the order issued by Major Gen. Lord Sterling during his command at the camp, and thanks him for the endeavor to preserve order and discipline, and the property of the farmers in the vicinity of the camp. He doubts not but the officers of every rank, from a just sense of the importance of securing to others the blessings they themselves are contending for, will use their utmost vigilance to maintain those privileges and prevent abuses, as nothing can redound more to their personal honor and the reputation of their respective corps. General orders, Alexander Scammil. Adj.-Gen

Directives were also issued from Regimental Headquarters:

All officers must see that their men attend upon prayers morning and evening and also at the service of the Lord's Day with arms and accoutrements ready to march in case of any alarm and that no drums be beaten until after the parson leaves the stage." [10]

The orders provided monetary penalties for not attending services. The funds collected from the offenses were donated to a fund created for sick soldiers of the company of the offender.

Observing Sunday as a day of worship was routine at Middlebrook and Pluckemin. There were always at least two services. Chaplains from the various regiments preached in rotation.[11] Services were conducted in churches near the camp, although there were only a few churches nearby in this remote rural area. Active congregations during the war were the Presbyterian Church at Bound Brook, the Hillsborough Reformed Church at Millstone, St. Paul's Lutheran Church in Pluckemin, and the Reformed Dutch Church of Raritan located in the Finderne section of Bridgewater. The parsonage of the Reformed Dutch Church was close to Washington's winter quarters in the Wallace House in Somerville, and he developed a warm friendship with his neighbor Reverend Hardenbergh. On October 27, 1779, the church was burned down by Lt. Col. Simcoe and his Queen's Rangers. All of these old houses of worship likely held services that were attended by both officers and enlisted men.

Log huts or clearings in the woods served as places of worship when there was no church nearby. Pulpits were set up on a rum barrel or a pile of knapsacks. Sunday began with prayers in preparation for the services which usually started at 11:00 a.m. The sermons offered to the troops seemed to have two basic themes. One presented all the attributes of the perfect soldier, encouraging the men to aspire to that model. The virtues of faithfully performing duties, cleanliness, honesty, temperance, and modesty were stressed and supported with biblical references. Other sermons portrayed the political and social issues of the day in an aura of patriotism and justification for the war.

New York Brigade Chaplain John Gano preached to a congregation that included men that had short enlistments who had not signed up for the duration of the war. He admonished them by stating: "I would never aver of the truth that our lord and Savior approved of all those who had engaged in his service for the whole war."[12] Despite his efforts, there is little evidence that Gano's rigorous preaching had any great influence in altering the behavior of the soldiers.

Sermons were long. Some continued for the entire day and required captive audiences of soldiers to attend in relays. Not all the men attended the sermons willingly. Those absent from worship were forced to spend equivalent time digging out stumps or other distasteful tasks. These measures ensured the presence of a large congregation.

Services were suspended during the periods when it was bitter cold, thus making it impossible to force the ragged troops to endure long sermons. Chaplains often took advantage of this pause in their activities to leave camp on furlough. In 1781 Reverend John Gano left the encampment at Pompton, New Jersey for the entire winter. When he returned, the pious Gano was greeted by a soldier who said, "We have been in want of everything during the winter, clothing, provisions and money, but hardest of all for us was that we did not have the word of God to comfort us." The conscience-stricken clergyman was distraught until he learned that the man was an incorrigible joker taunting him for having departed for the winter.[13]

We see rare evidence of religious diversification at Middlebrook. Presbyterians, Anglicans and other Protestant denominations worshipped together for the first time. This weakened the rigid lines between sects that existed in the men's home states. Only Catholics had their own services. Washington enforced toleration by issuing a directive that forbid anyone from publicly criticizing another religion. "Pope's Day," an unofficial holiday commemorated with the burning an effigy of the pontiff, was forbidden.

The Continental Army, with its corps of dedicated chaplains, always insisted that its soldiers remain virtuous. This policy may help explain why little evidence of sensual adventures can be found in soldier's diaries. There is even less evidence of immoral conduct in army documents. While sexual entertainment and attractions were often available, there were probably few common soldiers who could afford to participate.

The church and the clergy served as a positive force in the winter camps, providing care and comfort to the soldiers. Churches were often turned into hospitals, and through their sermons clergymen often showed empathy and acknowledged the adverse conditions in the camps. The men at Middlebrook in the winter of 1778-1779 endured physical suffering and privation, but there is ample evidence that even in the worst of times they were both reverent and creative as far as diversions, recreation, and amusement.

Chapter Thirteen

The Women of Middlebrook

THE WOMEN WHO accompanied the army ranged in personality from Martha Washington to prostitutes. They were never recruited or even encouraged to go along with the troops. Except for the rules and regulations that governed their behavior that can be found in camp records, few facts about the lives of the camp followers at Middlebrook are available to the historian. Their presence and numbers can be verified and interpolated from muster lists for food rations. Some details of their activities appear in other records from the war, but there is little recognition of their contributions and their critical role in the day to day life of the Continental Army. By the winter of 1777, around two thousand women marched with American troops and worked as seamstresses, nurses, and cooks. Their presence enhanced morale, improved the soldiers' welfare, and reinforced their will to fight.

General Washington was always reluctant to allow women to join his troops. In June 1777, he wrote, "the multitude of women in particular, especially those who are pregnant, or have children, are a clog upon every movement. The Commander-in-Chief therefore earnestly recommends it to the officers commanding brigades and corps, to use every reasonable method in their power to get rid of all such as are not absolutely necessary...."[1] When the Continental Army with its entourage of camp followers began marching to Middlebrook, Washington ominously warned, "The officers are to see that the Soldiers and Women who march with the baggage do not transgress the General Orders made for their government."

John U. Rees, Revolutionary War historian, examined several regimental records and found that, on average, adult female camp

The Women of Middlebrook

followers amounted to about three percent of the strength of a unit or one woman for every thirty men. Middlebrook far exceeded that number. This return, based on an army of about 10,000 men, shows 920 women at the camp in 1779:

Weekly return[s] of the Army under the Command of General Washington at Middlebrook N.J. and Park of Artillery at Pluckemin. 21 April to 28 May 1779:

Average Number of Women Per Company
(Nine companies per regiment)

Brigade	Regiments	Women/ Regiment	Women / Company
1st Pennsylvania	4	28	3
2nd Pennsylvania	4	27	3
1st Maryland	4	21	2
2nd Maryland	4	22	2
Muhlenberg's Virginia	4	11 (8 Companies)	2
Muhlenberg's Virginia	5	15	1
Woodford's Virginia	5	10 (8 Companies)	1
Woodford's Virginia	4	26	+
Scott's Virginia	5	17 (8 Companies)	1
Knox's Artillery		22 Companies	3

Who Were the Women Who Followed the Army?

The role of camp followers during the American Revolution has always been misunderstood. It is often assumed that most of them were prostitutes. True, a few desperate women may have turned to prostitution, but most did not. Of the thousands who accompanied the Continental Army, most were wives of the soldiers and officers. Many had children with them. They cooked, sewed, washed, mended, and cared for the sick and wounded.

The women often scavenged for food, clothing, and other supplies in local towns to help the soldiers survive in times of shortages. On campaigns, they hiked along with the baggage wagon train in the rear of marching columns. Their presence and ability to ease the tedious day to day burdens of the troops in the field, in in the dreary winter camps, and sometimes on the battlefield, must been an enormous boost to the quality of life for soldiers.

During the fiercest part of the Battle of Brandywine the women of the 6th Pennsylvania Regiment filled the empty canteens of soldiers under fire.[3] Another example is Molly Pitcher. This was a nickname given to a woman said to have been present at the Battle of Monmouth in June, 1778. She who is believed to have been Mary Ludwig Hays McCauley, the wife of American artilleryman William Hays, who was with Thomas Proctor's company, which was attached to the 4th Pennsylvania Regiment and billeted at the artillery camp. Many historians regard Molly Pitcher as folklore rather than a real historical figure, or suggest that she may be a composite image inspired by the actions of a number of real women. The nickname may have originated as a broad designation applied to women who carried water to men on the battlefield. Since the camp opened only six months after Monmouth, it is likely that the renowned Molly Pitcher lived at Pluckemin during the second encampment.

Washington soon learned that the act of driving women away would reduce the ranks of his volunteer army, and in the process he would lose some of his best men. He stated, "I was obliged to

Georgia's Nancy Hart repelling marauders. *(Library of Congress)*

give Provisions to the extra Women in these Regiments, or lose by Desertion, perhaps to the Enemy, some of the oldest and best Soldiers In the Service."[4]

Orderly books show that commanding officers tried to discourage the women from following the army. Over time they reluctantly recognized their efforts by first officially allotting them food. Each woman was authorized to receive a half ration each day "without whiskey." Children were allowed a quarter of a ration daily. Except for this food, the women were not compensated directly by the army in any other way for providing their important services. However, they did charge the men for performing chores, such as laundry service and sewing, and the army precisely regulated their prices.

Most regiments listed the names of all women and children along with the soldier to whom they "belonged." Unattached women did not receive rations and had to rely on the soldiers to share their food, or fend for themselves. Some of these women may have resorted to providing sexual favors to survive. In the end, the best chance a woman had to survive at Middlebrook was to be connected to a soldier.

Women and children were required to live outside of the campgrounds. Their soldier sponsors built shelters for them nearby and illicitly provided extra food to supplement their army rations. In general, while the women did provide vital services, they were always regarded as a nuisance by the army since they had to be fed and sheltered. The Quartermaster Departments were always searching for ways to limit their activities and obscure their presence. Women were accepted only if they were willing and able to work. A menacing entry in the orderly book of the 2nd Pennsylvania Regiment at Middlebrook in 1778 states, "Should any woman refuse to wash for a soldier at the rate [specified by the army] he must make a complaint to the officers commanding the company...who [if they] find it proceeds from laziness or any other improper excuse he is immediately to dismiss her from the regiment...no women shall draw rations unless they make use of the endeavors to keep the men clean."[5]

On April 21, 1779 it was reported that a large amount of rations had disappeared from Middlebrook stores. An annoyed General Washington suspected that the missing rations were being stolen by soldiers to provide for their women. He ordered an immediate investigation. "As the daily issues of provisions exceed, considerably, the total number of troops in this camp, I wish to know, on what days, in what manner, and by whose orders, these supplies were delivered."[6] This organized pilfering was impossible to control and continued until the camp closed in June.

When husbands fell in battle, wives suddenly found themselves alone, far from home and their relatives. To ensure that they did not starve, they quickly remarried, sometimes the next day and often to one of their husband's comrades. Women who had the misfortune of coming from Loyalist communities would be persecuted and rejected if they returned. For protection and security they were forced to remain with the army.

The women who followed the army to Middlebrook were not from any one social class. Martha Washington lived with her husband at the Wallace House in Somerville during the winter encampment. In fact, she joined the general every winter during the eight years of the war. She was a successful fund raiser for the impoverished army, and called on other women to donate money, clothing and supplies to the revolutionary cause. Her residence at the camps was the social center for visiting military leaders and foreign dignitaries. The efforts of "Lady Washington" ensured that adequate care was provided for wounded soldiers.

Martha Washington. *(The Miriam and Ira D. Wallach Division of art prints and photographs print collection, New York Public Library)*

Her presence and that of Catherine Greene, Lucy Knox and other wives of high-ranking officers at Middlebrook inspired the public and served as proof that everyone should make sacrifices for the cause for independence. Educated women, were typically officers' wives. They provided services such as copying, correspondence and the management of field hospitals.

Maria Cronkite was a typical example of the less affluent women who followed the army. In 1777, when she was thirty-two years old, she joined her husband, a fifer, in the 1st New York Regiment. "Mrs. Cronkite served as washerwoman for the officers until the close of the war when her husband was duly discharged...[and] had while in said service gave birth to several children."[7]

Slaves were among the camp followers at Middlebrook. An advertisement posted by the colonel of the 3rd Maryland Regiment in October, 1778 described a feisty and courageous woman who had run away from his unit. No other record has have been found regarding her life.

> *Mulatto slave, named Sarah, but since calls herself Rachael; she took her son with her, a Mulatto boy named Bob, about six years old, has a remarkable fair complexion, with flaxen hair: She is a lusty wench, about 34 years old, big with child; had on a striped linsey petticoat, linen jacket, flat shoes, a large white cloth cloak, and a blanket, but may change her dress, as she has other clothes with her. She was lately apprehended in the first Maryland regiment, where she pretends to have a husband, with whom she has been the principal part of this campaign, and has passed herself off as a free woman.*[8]

When the army left Middlebrook, women and children had to trudge nearly one hundred miles to the Hudson Highlands or Western New York State. They slogged for weeks over rough roads though the sweltering New Jersey wilderness. When they attempted to ride on the supply wagons they were forcibly removed by a direct order from General Greene, who complained that much needed wagons, "were being loaded with women."[9]

Army Nurses

Nursing was traditionally a female responsibility. Women could earn money and rations by becoming nurses with the Continental Army. The army preferred female nurses since they freed up men for fighting. Without the help of these women both the American and British Armies would have had to hire or assign men to these chores—and that meant diminishing ranks of soldiers.

Washington's request for one nurse for each 10 patients was authorized by the Continental Congress in 1775, but this was always an impossible goal at large winter encampments. At Valley Forge the sick and wounded numbered between 5,000 and 7,000 men. He lamented that more women did not participate and commented to General John Stark on August 5, 1778:

> *I cannot see why the soldier's women [camp followers] in Albany should be supported at public expense. They get most extravagant wages for any kind of work and are [fed]. [This is] robbing the public and encouraging idleness. If they would come down and attend as nurses to the hospitals they would find immediate employment.*[10]

Nursing generally meant caring for the troops by attempting to sanitize the sick and wounded and keep their wretched quarters as dirt free as possible. They washed patients, swept floors, and emptied chamber pots. Their efforts made an enormous contribution to sanitation during all years of the war. These women were not trained professionals; few could prescribe medicine or treat wounds though many performed those duties. They were usually hired from communities near to the larger encampments. They were authorized to be paid two dollars each month and were provided with full rations. Nursing was hard and even dangerous work and these valiant women were subjected to constant exposure to smallpox and other contagious diseases.

Redcoat Lasses

The Crown forces had a significantly higher number of women. In May 1777, the ratio of camp followers with the British forces in

New York was about one for every eight men. Later in the war, in August 1781, the troops in New York were shown to have a ratio of one woman to every four or five British soldiers.

The British Army officially accepted camp followers. Many men in the king's army were long service professionals and their families traveled with them to America on the same transport ships. The British found that allowing wives with families to accompany their husbands or partners provided a significant incentive to enlist. The average age of a Redcoat soldier was twenty-seven, so they were much more likely to have families than their Continental Army adversaries, whose average age was twenty. Many children were born to British and Hessian troops during the American campaigns. Somehow, these transient families seemed to thrive under adverse conditions of life in a foreign land.

The women with the British Army appear to have been more rebellious than their American counterparts. They often dropped out of the lines of march to plunder houses in nearby towns, and were later accused of bringing smallpox back from the quarantined homes they pillaged. Orders were issued to control this practice that specified that female offenders would suffer the same punishments as the soldiers

Camp followers Mary Colfritt and Elizabeth Clark were tried for plundering when the Crown Forces were marching through New Jersey from Philadelphia. A July 3, 1778 entry in the orderly book of a Maryland Loyalist regiment records that Mary Colfritt was found not guilty. Elizabeth Clark, on the other hand, was given no mercy. She was sentenced to "Receive 100 Lashes on her Bear Back and to be drummed out of the army in the Most Publick manner." Other entries show that women were often flogged for selling liquor to the troops.

General Washington was always conscious about the image of his troops, especially when they were viewed by the public. When the army marched through Philadelphia in 1777, he was humiliated by the ragged camp followers plodding along behind them, dragging pitiful children in tow. He ordered, "Not a woman belonging to the army is to be seen with the troops on their march thro' the city." But, after the parade of soldiers had passed, the defiant ladies appeared from out of the alleys where they had been hiding, and brashly paraded down the main streets of the nation's capital.

Chapter Fourteen

Festive Events and Interesting Visitors

Enliven Camp Routine

DURING THE REVOLUTIONARY War only a small fraction of a soldier's time was spent on the battlefield. At most, troops participated in one or two battles each year, and typical combat rarely lasted for more than a day or two. More time revolved around the struggles, hardships and monotonous routines of camp life. Ninety-seven men, about ten percent of the soldiers billeted at Middlebrook, were tried for crimes by court martial during the seven-month winter encampment. The most common offenses were absence without leave, theft, desertion and drunkenness.

General Washington constantly sought diversions and stimulating activities to relieve boredom and boost the spirits of his troops. He encouraged leniency when punishing offenders, and authorized dispensing a modest amount of liquor with daily rations. Holiday celebrations, and the welcoming of important or unusual visitors, brightened the camp during the grim winter days. These were events in which everyone could participate.

The Visit of the French and Spanish Ministers

A close alliance with France and Spain was established during the time the Continental Army was at Middlebrook. When the ministers of those countries visited the camp in early May 1779, it was a momentous event. France had formally recognized the United States on February 6, 1778. As a result, Britain declared war on France shortly thereafter, on March 17, 1778.

France and Spain had entered the war not out of overwhelming support for the interests of the Americans, but out of self-interest in their own wars with Great Britain. The Middlebrook meeting was critical. Its purpose was to negotiate he details of supporting the American cause and to plan a joint military strategy. An enthusiastic reception for the dignitaries was an opportunity for the struggling American nation to strengthen bonds with these countries and to request aid.

Preparations for the receptions of M. Girard of France and Don Juan Miralles of Spain were carefully planned, well in advance of their arrival in New Jersey. Four impeccably uniformed Maryland Battalions were selected to parade in front of the visitors. They were assigned to Drillmaster Baron von Steuben for preparation and training. The Prussian martinet relentlessly drilled them in precision maneuvers, and saluting with deafening barrages of artillery and musketry. The troops would demonstrate maneuvers of deployment, advancing, retreating and a bayonet charge. The event would conclude with a grand parade of thousands of men passing in review before the ministers.[1]

On the day the ministers arrived General Washington and other generals rode out to welcome them. They were apprehensive. It was expected that some regiments, inadequately funded by their individual states, would appear ragged and unkempt. In the end, the commanders were pleasantly surprised. General Greene wrote after the review, "We have just come from reviewing the troops, which made an elegant appearance and the French Ambassador and Don Juan, who were present, showed a great mark of approbation."

M. Girard was a guest of General Washington at the Wallace House in Somerville. After the welcoming ceremony, Baron von Steuben hosted a gala banquet at his headquarters at the Staats House in South Bound Brook. During the festivities the French minister

Reenactors portraying the 5th New York Regiment at Trenton. *(Author's collection)*

disclosed to a delighted General Washington that his country was prepared to immediately increase the flow of supplies and weapons to aid the patriot cause. Of even greater consequence was that Girard proposed that the powerful French fleet join the American forces and come under their direction.

The visit of the ministers to Middlebrook was acknowledged by all three countries to have been a great achievement. M. Girard complimented the Continental Army on its professionalism and predicted that his visit was the start of a successful long term American-French alliance.[2] France's entry into the war is regarded as a primary reason for America's success in winning independence.

A Procession of Fascinating Guests

The most welcome visitors of all to Middlebrook were family members. Most lived nearby and often visited the camps. At times they risked their lives to see a loved one if fighting was occurring nearby. Members of the Continental Congress sent to inspect the army, foreign diplomats, and even Native American chiefs also appeared at the encampment. Private David How wrote in his diary that "the King of the Ingans with five of his Nobles to attend him come to Head Quarters to Congratulate with his excellency."[3]

Festive Events and Interesting Visitors   159

In late May, a delegation of Delaware Indians stopped at Middlebrook on their way to meet with Congress. Unlike most of the tribes, they favored the American cause. They were warmly greeted by General Washington and were warned of the danger of opposing the Americans. Coincidentally, at the time he was planning the Sullivan-Clinton expedition to conquer and totally destroy the Iroquois Nations in western New York. The Native American representatives presented a memorial to Washington affirming their friendship for the American cause.

Governor William Livingston of New Jersey was a frequent guest. He had served as New Jersey's delegate to the Continental Congress from July 1774 to June 1776, had returned as brigadier generalof the New Jersey Militia, and was electedgovernor in 1776. During the war Livingston lived in Parsippany and Connecticut Farms. He traveled to Middlebrook several times to discuss the impact of military activities on the welfare of the civilians in his state.

In April, General John Sullivan arrived at the camp to start planning for the campaign against the Iroquois Nations. Native Americans overwhelmingly supported the British, since they were benevolent and did not encroach on their lands. Sullivan conferred with Washington and General Arthur St. Clair at the Wallace House. The invasion of New York would be launched from Middlebrook in July, and the offensive would last for several months. It was a major campaign of the war and involved almost a third of the American forces. The operation took the war to the homeland of the hostile Native American tribes in western New York State.

The punitive expedition was successful. Forty Iroquois villages and stores of winter crops were destroyed. The morale and the power of the six Iroquois Nations in New York was broken all the way to the Great Lakes. The wretched Indians fled to Fort Niagara, north of Buffalo, and also to Ontario, Canada to seek the protection of the British. With the military power of the Iroquois crushed, the vast Ohio Country, the Great Lakes regions, Western Pennsylvania, West Virginia and Kentucky were opened to post-war settlement.

General Benedict Arnold reluctantly left the comfort and lavish lifestyle of Philadelphia, as well as his Loyalist fiancée, the flirtatious teenager Peggy Shippen, to travel to Middlebrook. He arrived on April 20, 1779 in order to be put on trial. He was charged by the Pennsylvania Council for "appropriating public wagons of the state for private uses." The case was postponed several times. Arnold was irate and claimed that he was ill-treated and abused. He had a low opinion of the winter camp, writing to Peggy that he soon hoped to leave "these few dirty acres." He described Middlebrook as a "villainous plan, peopled by villainous men."[4]

At the time, General Arnold had the support of all high ranking officers, including Washington. He was remembered for his former heroic exploits as the nation's most daring and brilliant general. An empathetic General Knox wrote to his brother from Middlebrook, "You will see in the papers some highly colored charges against General Arnold by the state of Pennsylvania. I shall be exceedingly mistaken if one of them can be proven. He has returned to Philadelphia and will, I hope, be able to vindicate himself from the aspersions of his enemies."[5] The charges against Arnold were dropped. Six months later he defected to the British and became the most notorious traitor of the war.

The Wretched Convention Army

Taunting the unfortunate bands of prisoners of war and captive Tories that passed through the camp provided endless amusement for jaded soldiers. The "Convention Army" was made up of 5,900 British prisoners of war captured at the American victory at Saratoga in 1777. The pathetic procession plodded wearily through the Middlebrook camp on a 700 mile march to captivity in Virginia. They pleaded with the Quartermaster Department to provide food for the sick and wounded among them. In October 1778, a delegation of their officers appealed directly to General Stirling for relief. At the time American supplies were dwindling and nothing could be done to ease their suffering. The prisoners were treated with empathy and respect, but assistance could not be offered, and they continued their tragic trek to Virginia.

Holidays - The Social Highlights of Life at Middlebrook

Holiday celebrations were one the few organized diversions for the common soldiers. Official holidays were Christmas, Thanksgiving, Fourth of July, May Day, Burgoyne's surrender at Saratoga and the King's Damnation Day on September 22, a date that had been celebrated before the war as the King's Coronation Day. The commemoration of the French Alliance was added at Middlebrook in February 1778. Benevolent local civilian often contributed a supply of rum to enhance the revelry.

Celebrations usually consisted of a parade, a sermon by a chaplain, extra rations for the privates and a dance for the officers. Thanksgiving and Christmas were always observed with religious services, light duty, and an extra allowance of liquor. Washington issued this proclamation for Thanksgiving a year earlier at Valley Forge:

Tomorrow being the day set apart by the Honorable Congress for public Thanksgiving and Praise; and duty calling us devoutly to express our grateful acknowledgements to God for the manifold blessings he has granted us. The General directs that the army remains in its present quarters, and that the Chaplains perform divine service with their several Corps and brigades. And earnestly exhorts, all officers and soldiers, whose absence is not indispensably necessary, to attend with reverence the solemnities of the day.[6]

The last troops departed from Middlebrook before the 4th of July 1779, so that holiday was not celebrated at the camp. It was considered the premier event of the year, and was zealously observed by all units of the Continental Army during the duration of the war. On that holiday, the entire army was paraded, and massed artillery fired thirteen thundering salvos, one for each of the colonies. This was followed by running volleys of musketry. After a rousing speech by a senior officer, the ceremony ended with three huzzas from the thousands of troops. Routines were relaxed for the rest of the day and a gill of rum was issued to each man.

The sound of gunfire salvos at the celebration in 1778, in New Brunswick, New Jersey, were heard twenty-five miles away on Sandy Hook. The evacuating British panicked. They believed that they were being attacked. Elijah Fisher a member of Washington's Life Guard, described this July 4, 1778 event in his journal, " We celebrated the Independence of America the whole army paraded and at the right of every brigade there was a field peace[cannon] placed, then was the signal given for the howl arm to fire and they fired one round a piece and the artillery discharged thirteen cannon and we gave three cheers, At night his excellency [General Washington] and the gentlemen and ladies had a bawl at head Quarters with great pomp."[7]

General orders were issued on March 16, 1780 proclaiming St. Patrick's Day as an official holiday, in order to honor the large number of immigrants and deserters of Irish descent from the British Army. People of Scots-Irish descent were the largest immigrant group to arrive in the colonies during the 1700s. Estimates are that more than one quarter of the Continental Army was Irish by birth or ancestry and nearly one half of the men in the Pennsylvania and Maryland regiments were Irish. Many of these early immigrants were Presbyterians from the northern county of Ulster who had settled in the Appalachian uplands. They collectively referred to themselves simply as "Irish." This occasionally causes confusion since the Irish are more often historically identified with the great wave of Irish immigration that occurred during the potato famine of the 1840s and 1850s.

General Washington devised another creative way to provide diversion and improve morale. To provide more leisure activity during the dreary and uneventful days of the long winter at Middlebrook, he declared certain days "religious holidays," or "Continental Thanksgiving" days. The first one was held on December 30, 1778, during a severe cold spell. The entire army was given a day without work or other activities. On these days Washington requested that soldiers give thanksgiving and praise for the good that had come to the new nation since the start of the war. Regimental chaplains carried over that theme into their sermons. Troops were paraded before the services. Dr. Thacher commented, "This day of rest was appreciated by everyone and was repeated again on May 6, 1779."[8]

Rum, Whiskey, Brandy, and Home Brew

After the huts had been completed, the restless men at Middlebrook had more idle time and desperately sought diversion. General Washington had banned the sale of liquor in camp when the troops began arriving. But, while destitute soldiers often went without adequate food and clothing, somehow liquor always seemed to be available. Court martial summaries regularly mention drinking infractions: "getting drunk and behaving in a disorderly manner."

Drunkenness was a common offense in camp. Liquor, although not distributed by the army as rations, was readily available from the suttlers. These civilian merchants with portable shops near camps could always provide liquor for those who could afford it. They could also sell it to those without money as an advance on their pay. Gullible soldiers could drink up their entire pay before they received it if they weren't cautious. At other times soldiers traded their blankets and clothing for alcohol. Later in the war, regulations for suttlers stipulated, "For liquors or other articles sold to noncommissioned officers & soldiers, artificers and waggoners, nothing shall be taken in payment but money."

Soldiers also frequented taverns near camp. Steeles Tavern, adjacent to the Virginia Camp site on today's Steeles Gap Road, near

Foothill Road in Bridgewater, was such a popular stop that it was officially declared off limits. On January 1, 1779, General Smallwood ordered, "tippling at houses in the vicinity of the camp is forbidden." The taverns continued to flourish for the entire duration of the encampment despite being declared illegal.

References to liquor appear on official requests for supplies and in general orders. The official distribution of rum, whiskey, hard cider and brandy was especially liberal at the Pluckemin Artillery Camp. General Henry Knox insisted that quartermasters provide daily rations: "We have found by experience that this would support the men through every difficulty." The accepted belief at the time was that drinking in moderation promoted health and reduced fatigue.

Alcoholic drinks were used to celebrate most special occasions. They were often issued after a day of extra work, a job well done, after a long march, and during bad weather. The standard allowance per day was a gill (one half pint) which was often mixed with three parts of water and carried in canteens. A gill was just enough to induce mild euphoria, but not enough to get men drunk. In addition to imbibing on a regular basis while in camp, the Continental Army, and other armies of the world at that time, issued copious amounts of spirits before and during battles.

Party Time for Officers

Officers at Middlebrook lived in better quarters and had more fulfilling leisure activities than the soldiers. Officers were often entertained in homes in the vicinity of the camp. Diaries and other accounts are replete with vivid descriptions of frequent social events that were enthusiastically attended by the generals, as well as lower ranked officers.

Dr. James Thacher, the twenty-four-year-old Massachusetts surgeon, celebrated New Year's Eve at a party given by the Virginia Regiment. He wrote in his diary, "After a dinner at Colonel Gibson's the celebration moved to General Muhlenberg's quarters where the dancing continued until three in the morning." During the holiday season Thacher "did not have a day without receiving invitations to dine, nor a night without amusement and dancing."[9]

Major Robert Forsythe, aide to General Greene at Middlebrook, attended a party of fellow officers and unattached local ladies on January 27, 1779: "We had a most agreeable hop last evening-present were Mrs. William Livingston, Miss Cornelia Lotts, Miss Harriet Van Horn, Miss Betsy Livingston, Cols. Biddle, Livingston, Butler, Williams and Scammell were there... we kept it up until four in the morning."[10] Jeduthan Baldwin, a lieutenant of the artificers, held a house warming " to celebrate the completion of his hut."[11] General Greene commented that General Washington danced three hours with his wife without sitting down.[12] Many unmarried daughters of the more affluent local farmers were present at the parties. The usual round of pleasure for the officers included dancing, dinners, teas, sleighing parties and horse riding parties.

The greatest social event of all during that winter was the commemoration of the Alliance with France, celebrated at Pluckemin on February 18, 1779. A grand ball was held at the Academy there on February 18, 1779 and was attended by 400 officers and civilian gentlemen, and about 70 ladies.

Wives of high ranking officers joined their husbands as soon as the army arrived at the Middlebrook winter quarters. Martha Washington, Catherine "Kitty" Greene, and Lucy Knox made it a practice to spend the winters with their husbands. Mrs. Washington was in the habit of saying that she always heard the last cannon fired in the fall and the first one in the spring. The arrival of the wives signaled the beginning of festivities among officers and families.

Dinner was the favorite method of entertaining. "General Greene and his lady present their compliments to Colonel Knox and his lady and should be glad for their company tomorrow at dinner at two o'clock." Often the dinners were meager. Officers and privates suffered alike when food was scarce, but the social events did not rely on the quality of food. One such dinner is described as having been only potatoes with beechnuts for dessert.

While occupying the Van Veghten House in Bound Brook, General Greene commented on March 19, 1779 that many officers and their families traveled to Trenton to meet friends from Philadelphia. Hospitality offered by civilian gentry in the area also kept officers enjoyably entertained in their off duty hours.

There was an active social life for everyone at Middlebrook. Hospitality offered in the homes of senior officers and civilian gentry in the area kept officers enjoyably entertained during their off-duty hours. In fact, the end of the encampment ended a number of courtships. The enlisted men made their own fun in camp or in nearby taverns. Even General Washington knew how to relax and have a good time.

Chapter Fifteen

Crime and Punishment

A VAST NUMBER of court martial trials appear in orderly books, the daily records kept during the Middlebrook Encampment. These documents portray a litany of crimes and punishments. They show that ninety-seven men, about ten percent of all soldiers at the camp, were tried for crimes by court martial during the winter at Middlebrook. There are repetitive entries for desertion, insubordination, theft and destruction of civilian property, and drunkenness. Most sentences involved some method of corporal punishment. Soldiers charged with minor offenses were tried by a regimental or garrison court consisting of five officers. More serious offenses were tried by a general court martial made up of thirteen officers.

Robbery was the most common crime at Middlebrook. The usual penalty for this violation was to imprison the offender in chains, and put them on a diet of bread and water. Flogging, varying in severity from fifty to a hundred lashes, was another option for this crime.[1] On June 1, 1778, Private William Scully was tried by court martial for robbing the house of civilian Robert Dennins and stabbing William Cox with his bayonet. He was sentenced to receive 100 lashes to be applied in view of his regiment, the 1st Virginia.[2] In April, a sergeant and six privates of the Light Infantry Corps were tried for entering and robbing three houses near the camp. Their punishment was to receive 100 lashes each.[3]

The most serious offense that could be committed at Middlebrook was desertion. The penalty for desertion of any duration was always death. On April 14, 1779, three men of the 1st Maryland Regiment were tried for desertion and were sentenced to death by hanging on April 20, 1779. Two other men who previously received the death sentence were also scheduled to be hanged that day. On the morning the entire camp was gathered to witness the execution. The five men, each of which was sitting on his own coffin with a noose around his neck, were transported to the gallows in a cart. Upon their arrival, a messenger rushed up with the astonishing news that General Washington had pardoned three of the criminals. The others were executed.[4] This hanging took place in Pluckemin, which may indicate that the gallows for all of the Middlebrook camp were located there. This is the only specific reference to a gallows location in the camp that has been found.

Crime at Middlebrook was not limited to enlisted men. Officers, however, received significantly lighter sentences. This blatant discrimination and favoritism, unquestioned and codified under military law, was modeled on the British Articles of War. Soldiers received physical punishment, while officers who engaged in "ungentlemanlike behavior" were punished by reprimands or humiliation. For an "officer gentleman" of that time, public humiliation in front of the army or civilians, or discharge from the army, was considered a severe punishment.

Hugh Smith, an army postmaster, was tried on April 5, 1779 for inciting a riot that involved a number of soldiers. Smith was the judge at a cockfight, a common event at the camp. He made a ruling that was disputed and afterwards attacked the challenger with his sword. As a result, he was forced to resign his commission and was discharged from the Army. A Commissary Department officer named Lewes was charged with embezzlement and selling public stores. He was acquitted, but General Washington reversed the decision after Lewes admitted to purchasing liquor from civilian merchants and reselling it to soldiers for a profit. He too was discharged from the Army.[5]

Army regulations stated that any soldier who was arrested or confined had to be brought to trial within eight days or as soon as a court could convene. During campaigns that lasted several months it was difficult to convene court martials and obtain prompt hearings, since the army was constantly on the move. General Washington insisted that were postponed trials be held without delay as soon as his forces settled into a camp. This created a backlog of cases and explains why such an excessive number of court martial hearings and sentences appear in the orderly books at the Middletown Encampment.

While George Washington's image is that of a benevolent military leader who was obsessed with the welfare of his adoring troops, he firmly supported all types of corporal punishment, including the death penalty. A dilemma for American military leaders at that time was that the existing military legal system did not provide for any degrees of punishment between the maximum 100 lashes and death. Although 100 lashes seem to be an extremely cruel method of punishment, this penalty stood next to death. As a result, when court martials believed a punishment warranted more than 100 lashes, they were forced to go to the next level and hand down a death sentence. This action resulted in an inordinate amount of death penalties.

The 100-lash limit was faithfully observed during most of the war, although Washington regarded it as inadequate for many crimes.[6] After the Pennsylvania Line mutinied at Jockey Hollow in 1781, he insisted that the existing 100-lash limit be raised to five hundred. His demand met with strong opposition from Congress. Punishment limits were set by Congressional regulations, and even the Commander-In-Chief was not given the power to authorize an increase in severity.

Washington tempered his view on sentences by always insisted on restraint, "lest the frequency of punishment should take off the good effects intended by it." He warned his generals "not [to] introduce capital executions too frequently," and reminded them that if executions were too common they would "lose their intended force and rather bear the Appearance of cruelty than Justice."[7]

American commanders were constantly attempting to balance harshness and leniency by trying to define levels of punishment that would enforce discipline but at the same time avoid provoking mass desertions or mutinies. While they were convinced that corporal punishment was the way to ensure absolute obedience and respect for rank, they knew that it could inspire rebellion if the troops regarded punishments as being too severe. Harsh sentences had already contributed to inspiring dangerous insurrections.

In New Jersey camps soldiers became mutinous, made threats, and openly voiced their discontent over the pattern of cruel and insensitive sentences. The 1780-1781 mutinies at Morristown and Pompton of Connecticut, Pennsylvania, and New Jersey Regiments, which took place a year after Middlebrook, were in part caused by resentment over severe sentences. There is no other evidence at the camp that indicates that frustrated and terrified officers believed there was an imminent mutiny.

In real practice the death penalty was frequently reprieved, and many sentences were never carried out. Only an estimated 17% to 30% of capital sentences actually resulted in executions. The reason for frequent pardons of death sentences was pragmatic—executions resulted in the loss of manpower to the army. Each death meant the loss of a fighting man; flogged offenders could recover and live to fight on.

The army also recognized that if the word spread that callous punishments were inflicted for relatively minor violations, men would more readily desert. Stories brought home by returning soldiers and deserters of men punished for the slightest offenses also had a chilling effect on recruiting. Reports of cruel and arrogant officers who were handing out sentences of 50 lashes for cutting up a blanket or urinating near huts discouraged volunteers. Clemency appears to have occurred when leaders feared repercussions. Mercy, while frequent, was not altruistic.

Capital punishment was far more common in the American Revolution than in all the U.S. wars that followed. Desertion, spying, and mutiny were considered the worst crimes under the Continental Articles of June 30, 1775. These offenses carried a mandatory death penalty by either hanging or firing squad. Executions were usually performed with the largest audience of soldiers possible. The tragic display of this most extreme form of military punishment was meant to terrify anyone contemplating committing a criminal act.

In his diary, Army Surgeon James Thacher depicts the tragic punishment of men of the New Jersey Brigade who mutinied on January 20, 1781 at Federal Hill, near Pompton. The site today is now adjacent to Route 287, in Bloomingdale, New Jersey. The men had not been paid in over a year and lacked warm clothing. Most suffered from frostbite and scurvy. Many of them had enlisted for "three years or the duration of the war" and assumed that these terms meant "whichever comes first." They soon learned that they would not be going home, and were compelled to continue serving for the duration of the war. Neglected by both military leaders and the Continental Congress, these men felt that they could not endure another winter. After the rebellion was suppressed, three men were identified as the ringleaders and sentenced to be executed by firing squad.

Twelve of the most guilty mutineers, were next selected to be their executioners. This was a most painful task, being themselves guilty, they were greatly distressed with the duty imposed on them, and when ordered to load, some of them shed tears. The wretched victims, overwhelmed by the terrors of death, had neither time nor power to implore the mercy and forgiveness of their God, and such was their agonizing condition, that no heart could refrain from the emotions of sympathy and compassion.

The first that suffered, was a sergeant, and an old offender, he was led a few yards distance and placed on his knees; six of the executioners, at the signal given by an officer, fired, three aiming at the head and three at the breast, the other six reserving their fire in order to dispatch the victim, should the first fire fail; it so happened in this instance; the remaining six then fired and life was instantly extinguished.

The second criminal was, by the first fire, sent into eternity in an instant. The third being less criminal, by the recommendation of his officers, to his unspeakable joy, received a pardon. This tragical scene produced a dreadful shock, and a salutary effect on the minds of the guilty soldiers. Never were men more completely humbled and penitent; tears of sorrow, and of joy, rushed from their eyes, and each one appeared to congratulate himself, that his forfeited life had been spared.[8]

Surgeon Thacher also vividly describes a scheduled mass execution that occurred at Jockey Hollow in May 1780. Eleven men were scheduled to be executed, all but one for desertion. Their graves were dug, and eight of the men were standing on ladders with ropes around their necks. Thatcher portrays the scene:

This was a most solemn and affecting scene, capable of torturing the feelings, even of the most callous breast. The wretched criminals were brought in carts to the place of execution. Mr. Rogers, the chaplain, attended them to the gallows, addressed them in a very pathetic manner, impressing on their minds the heinousness of their crimes, the justice of their sentence, and the high importance of a preparation for death. The criminals were placed side by side, on the scaffold, with halters round their necks, their coffins before their eyes, their graves open to their view, and thousands of spectators bemoaning their awful doom. The moment approaches when every eye is fixed in expectation of beholding the agonies of death, the eyes of the victims are already closed from the light of this world. At this awful moment, while their fervent prayers are ascending to Heaven, an officer comes forward and reads a reprieve for seven of them, by the Commander in Chief.[9]

In the end, General Washington finally reprieved ten of the men. One man, James Coleman, was executed. He was considered the worst of the offenders because he had forged discharges that had enabled more than one hundred soldiers, including himself, to leave the army. Reprieves were held back until the last possible moment to maintain a high level of apprehension among the spectators. Executions were carried out with enough frequency that anxiety and doubt always existed in the ranks.

Less brutal penalties were imposed for petty crimes. In these cases, offenders were often fined. A month's pay was the maximum permitted by the Continental Articles. Orderly books are replete with sentences of fines for minor infractions. Confinement on bread and water for lesser offences was also a common sentence for enlisted men. Other punishments were created to publicly disgrace and degrade offenders. Men were forced to wear a signs which named their offense, tied naked to a tree in a public place, made to wear women's clothes, or chained to a heavy log that had to be dragged wherever they went.

Both men and officers were "drummed out" of the army, an action meant to humiliate them. They were followed through camp by drummers and fifers and ejected at the front entrance. This practice was continued in the U.S. Marine Corps until 1962.[10] Noncommissioned officers frequently were reduced to ranks for their crimes. A jail probably existed in nearby Raritan where John Beatty, Commissary General of Prisoners, had his headquarters.

The Cat Out of the Bag-Frequency of Flogging at Middlebrook

Flogging was the usual punishment for many trivial offenses. Stealing other soldiers' socks, urinating in the woods, dozing on watch— all resulted in lashes. Although the number of stripes should have been related to the severity of the offense, in reality most offenders received the maximum of 100 lashes for all types of transgressions. The brutal task of flogging was performed by drummers and fifers under the direction of the regimental drum-major.

The offender was tied to a tree or post and in most cases was required to bite down on a piece of lead. The whip was formed from several small knotted cords that cut through the skin of the naked back at every stroke. The stinging blows were usually absorbed in silence. To increase the severity of the penalty, the prescribed number of stripes was often administered in installments so the flesh of the victim could heal partially and became inflamed, at which poinit the flogging then resumed.

Surgeon Thacher described the cruel practice of flogging in 1780. He was impressed by the bravery and defiance of the soldiers subjected to this brutality.

> *However strange it may appear, a soldier will often receive the severest stripes without uttering a groan, or once shrinking from the lash, even while the blood flows freely from his lacerated wounds. This must be ascribed to stubbornness or pride. They have, however, adopted a method which they say mitigates the anguish in some measure, it is by putting between the teeth a leaden bullet, on which they chew while under the lash, till it is made quite flat and jagged. In some instances of incorrigible villains, it is adjudged by the court that the culprit receives his punishment at several different times, a certain number of stripes repeated at intervals of two or three days, in which case the wounds are eating of inflammation, and the skin rendered more sensibly tender; and the terror of the punishment is greatly aggravated.*[11]

While the punishments inflicted by the Continental Army seem barbarous, British Army penalties were even more severe. Sentences of 500 to 1,000 lashes amounted to a death sentence. In 1803 the maximum number of lashes that could be imposed even reached 1,200, which always permanently disabled or killed the offender. This maximum sentence was only inflicted nine or ten times by general court martial, but sentences of up to 1,000 lashes were administered more frequently. In the British Army and Navy these brutal punishments were often inflicted for trivial offenses. One soldier was sentenced to 700 lashes for stealing a beehive. These draconian measures were abandoned in the early 18th century, when both military and civilian offenders were instead shipped to the penal colony in Australia. Once there, more whippings often awaited them.

The doling out of punishments was apparently personally distressing for General Washington. He begged his men to save him from the anguish of approving punishments. His daily personal life in service was comfortable, which may explain why he was mystified by the behavior of soldiers. In a rare moment of insensitivity he complained, "Why will Soldiers force down punishment upon their own heads? Why will they not be satisfied to do their duty, and reap the benefits of it? Why will they abandon, or betray so great a trust? Why will they madly turn their backs upon glory, freedom and happiness?"[12] Though he favored administering severe penalties he always preferred to have "the business of the Army conducted without punishment."

Chapter Sixteen

Supporting a Deprived Army

THE LACK OF an established supply system blocked efforts of the Continental Congress to equip and feed its army. The new government had never been confronted with such an immense and complex task and it was not prepared to administer an efficiently organized system. Efforts to provide for the army were generally short-term, incompetent, and intensified by lack of funding and transportation. General Nathanael Greene had the challenging task of serving as the Quartermaster General during the difficult months at Middlebrook.

He often lamented the never-ending shortage of land transport. There were never enough arts, wagons, sleighs, sledges, oxen, horses, harness, packs, or saddles. Forage was scarce and wagoners and carters were difficult to recruit. The situation was so dire that the Camp Committee at Middlebrook wrote to Congress on February 12, 1778, "almost every species of camp transportation is now performed by men, who without a murmur, patiently yoke themselves to little carriages of their own making, or load their wood and provisions on their backs."

Support and maintenance facilities at Middlebrook were dispersed over many miles. They were typically located near roads or waterways, and were often close to the living quarters of the commanding officer responsible for the service. General Greene lived at the Van Veghten House; quartermaster's stores were located opposite the house along what is today Van Veghten Drive in the Finderne section of Bridgewater. Equipment was also fabricated and repaired in Finderne at the artificers shop on Old York Road. The post office for the camp was on the property of what is now the Somerset County Court House in Somerville. The horse corral was on today's Route 202/206 in Bridgewater, about a mile south of Pluckemin, while slaughter pens were at Bound Brook.

Support Facilities

Artificers Shop

Artificers were skilled artisans and mechanics who kept military equipment in good working order. George Washington established a Corps of Artificers as part of the Continental Army to perform all trades that supported the military. They included blacksmiths, wheelwrights, tailors, and craftsmen who repaired pistols and muskets. Small foundries and iron works fabricated parts and cast buckles, buttons, and musket balls. Tents were also brought here for repair. The artificers provided the army's ammunition, repaired all wagons, and patched up small arms and tools. At one time, General Greene commented that an army without artificers is like a man without hands.

The artificers camp at Middlebrook was under the command of Colonel Jeduthan Baldwin. In February 1779 Baldwin's regiment was staffed with 186 artificers. Their operations were located in four wood frame buildings on the north side of Old York Road, about a mile west of the Van Horne House. In terms of today's streets, this area would cover Grand Boulevard north of Somerset Street and Oak and Pine Streets in Somerville.[1]

Artificer. *(Courtesy N.W. Canton and Pamela Patrick White, New Windsor Cantonment)*

Winter Encampment at Middlebrook. *(Courtesy Harpers Ferry Center Commissioned Art Collection, National Park Service)*

Clothier's Office

Clothing was scarce during the encampment, until apparel began arriving from France in the spring. Clothiers issued these new uniforms to the regimental quartermasters for distribution to the troops. Their facility was located several miles west of the main camp, along today's Route 202/206, near Pluckemin. Robert Erskine's map places it near the Ten Eyck house at the junction of the Lamington River and the north branch of the Raritan River.[2] Shoes were made off-site at a factory in Newark that was staffed by troops from the Maryland Regiments.

Post Office

The mail depot for the Middlebrook Encampment was on the north side of Old York Road near Tunison's Tavern, about a half mile from Washington's Headquarters at the Wallace house in Somerville.[3] The tavern, built in 1748, was owned by Cornelius Tunison, and stood on the corner of Main Street and Grove Streets. The postmaster during the winter encampment of 1778-1779 was Hugh Smith.[4] Troops were also billeted at Tunison's during that encampment. In 1781, Somerset

County Court was moved from Millstone into the tavern; the tavern and the Somerset Hotel that replaced it served as a stagecoach stop for many years. It was the oldest continuing business in Somerville. The hotel still stands on the site today.

While mail deliveries were often disrupted by the war, the postal system in the country was well established at the time of the Middlebrook Encampment. Benjamin Franklin, formerly the postmaster of Philadelphia, became Postmaster General for the colonies in 1753. In 1774, the British fired Franklin from his postmaster job because of his pro-independence activism. When the country declared independence in 1776, he was appointed Postmaster General of the United Colonies by the Continental Congress. As was his nature, he made many improvements to the mail system on colonial routes. He succeeded in cutting delivery time in half between Philadelphia and New York by having the weekly mail coach travel day and night with relay teams. Franklin also standardized delivery costs based on distance and weight by introducing the first-rate chart. He left an improved mail system with routes from Florida to Maine before he was sent to France in 1776 as a diplomat.

Quartermasters Stores

Equipment used to set up and maintain the facilities of the camp itself was called quartermaster stores. This included tents, tools and utensils. This gear was stockpiled in Morristown, where new accessories were fabricated and used material was salvaged. Tents that could not be reused were made into wagon covers, forage bags, knapsacks and even canteens. Some of this equipment and tools had been shipped to Morristown directly from Valley Forge.

The quartermaster's depot at Middlebrook was located opposite the Van Veghten house, the temporary home of General Nathanael Greene.[5] The army requisitioned two houses there to store the equipment. Greene had reluctantly accepted the quartermaster role, a noncombatant office, in March 1778, and he held it until August 1780. Most of his tenure at Middlebrook occurred during a time of nagging shortages and transportation difficulties.

Commissary Department

The lack of cash or acceptable credit made it very difficult for the Army Commissary agents to obtain enough food to sustain the troops. The Commissary Department, managed by Royal Flint, Assistant Commissary General of Purchases Commissary, was headquartered in Raritan.[6] Colonel Jeremiah Wadsworth, Commissary General of the Continental Army, advanced his personal funds to purchase food when supplies at the camp became depleted. He resigned soon after the army left Middlebrook, after being notified that Congress would not be able to reimburse him.

On November 4, 1775 Congress specified that daily rations were to be provided for the soldiers of the Continental Army. This directive was in effect at Middlebrook and did not change during the course of the entire war. It was very similar to that which was stipulated for British soldiers.[7]

> *Resolved, that a ration consists of the following kind and a quantity of provisions, viz:*
>
> *1 lb. of beef, or 3/4 pork, or 1 lb. salt fish, per day*
> *1 lb. of bread or flour per day*
> *3 pints of pease or beans per week, or vegetables equivalent*
> *1 pint of milk per day*
> *1 half pint rice or indian meal per week*
> *1 quart of spruce beer or cyder per day*

Also listed were molasses, candles, and soap, all of which was to be issued on a company basis.

For the most part, food for the troops consisted of two basic items—bread and meat (beef or pork). This constant diet of half-cooked meat and hard bread was not healthy, as General Washington was well aware. He often lamented the lack of vegetables, vinegar, or beverages to provide vitamins which might help prevent scurvy.

No archeological evidence of field kitchens or other centralized food services has been found at the Middlebrook camp. The men prepared their own meals at their hut sites. The only utensil issued to troops at Middlebrook was a camp kettle—one for every six men. These had a capacity of nine quarts and were used to boil beef or other meat. Vegetables, when available, were added to the pot to make a stew.

All food was prepared by the soldiers themselves, with the exception of bread. A regimental commander would dispense the flour supply of his unit to army bakers or civilian cooks. A pound of flour was drawn from the commissary for each soldier. This flour produced far more than a pound of bread when water and other ingredients were added, so the surplus bread could then be sold by commissary officers who could purchase other supplies with the earnings. While at Pluckemin, General Knox wrote to General Washington that his commissary made so much profit from selling excess bread that they were able to purchase 8,000 rations from local farmers for the Artillery Park.

In early 1778, a seventy-five-man company of bakers was set up to keep the army supplied with fresh bread. Bread was produced in field ovens constructed in the camp, or in the ovens of bakers in nearby towns. It was estimated that it took approximately one barrel of flour (about sixty gallons), to provide bread for each soldier during the seven-month campaign season. Flour supplies in the northern and middle colonies gradually diminished during the encampment. New England was not a large grain producer and supplies in the middle colonies were soon exhausted. Consequently, the army relied more on the South as the war went on.

The necessity to pay in cash for all army provisions, whether purchased or seized from civilians, created problems. Over time, the currency and credit of Congress and the army became virtually worthless. The existing commissary collection system was discontinued by the time of the arrival of troops at Middlebrook in late 1778. A new plan was devised that required each state to provide specific quantities of flour, corn, beef, pork, rum, salt, and tobacco to the Continental Army.

This process was not successful, despite desperate appeals issued by Congress and from Middlebrook to each state. The arrival of armaments and other supplies from France after the army left Middlebrook in 1780 did not relieve food shortages. Ironically, the French army, with thousands of troops in the northeast, began competing for the same provisions.

Slaughter House

The slaughter house and animal pens were located in the center of Bound Brook. This site was likely chosen to be near to the homes of the many civilian butchers who worked there.[8] Cattle were driven from other states to be slaughtered at this central location. It was managed by a civilian, William Jones, as were the other support facilities of the camp.[9]

The process for obtaining fresh meat was to purchase cattle and hogs wherever available and drive them to the central pasturage at Bound Brook for butchering. An alternative was to slaughter the animals where they were purchased, salt the meat, and then stow it in barrels for shipment to Middlebrook. These preserved rations were essential when the army was in the field, since herds of animals could not be safely moved along with the troops. Centralized stores of salted meat were stockpiled along the New Jersey routes of march. One of these reserves was at Bound Brook.

Horse Corral

The number of horses allowed at Middlebrook was strictly limited, due to the scarcity of forage in the Raritan Valley. Only animals necessary for the operation of the camp were permitted. The horse corral was located about two miles below Pluckemin. The facility was situated a quarter of a mile below Harding Road where Route 206 turns into Somerville Road. John Church, a civilian manager, was in charge of the corral. A forage silo was built in Raritan.[10]

Hospitals

Two hospitals served the needs of the Middlebrook Encampment. One was located at Somerset Courthouse, now Millstone. The main army medical facility, managed by Dr. George Draper, occupied three barns there.[11] Another hospital was in New Brunswick. Additional medical facilities were in Bound Brook near Drapers's quarters. Most of the sick were cared for by surgeons within the camp at a number of huts set aside in a remote section of each regimental campground. No records have been found that show the number of sick in camp or those transferred the hospitals. While the quantity of men ill at the winter encampment of 1778-1779 is unknown, it must have been substantial. A return at Middlebrook in May 1777 shows about thirty percent of the troops—2,660 out of 8,298—were unfit to fight. Most of them were likely sick or disabled.

When the artillery regiment began leaving in the spring of 1779, General Washington ordered that the sick and wounded housed in barns and other public buildings at Middlebrook be moved to the vacant, well-constructed buildings at Pluckemin. The location soon became one of the army's three general hospitals in New Jersey, with more than 1,100 patients.

The army was crippled by disease during nearly every campaign for the remaining years of the war. At the time of the Middlebrook

Encampment only nine permanent army hospitals existed in the entire northeast. Death rates were high in these primitive facilities due to close quarters, poor sanitation, and contaminated food and water supplies. Only small amounts of medical equipment were available.

As the command center of the Continental Army, Middlebrook was responsible for all units of the army, anywhere in the field. In addition to supporting the troops at the main camp at Middlebrook, there were requirements at other locations. Regiments at Pluckemin and Paramus had to be supplied, as well as the many sick and wounded scattered in field hospitals in other states.

Feeding a Hungry Army

As much as a third of all food for the Middlebrook Camp came from the surrounding Raritan and Washington Valleys. Somerset County was a rich agricultural area, which made it an important consideration for the selection of the campsite. Wheat, hay, beef, and vegetables for the camp were all purchased from local merchants whenever possible. Foraging expeditions were regularly sent out from the camp. Food was typically sold voluntarily by Patriot farmers; however, many claims for damages, filed by farmers after the war, provide ample evidence that force was often used if it became necessary. Known Tory farms were consistent pilfering targets.

Other food sources for the Middlebrook Encampment were in distant locations. Cattle came from New England and eastern Pennsylvania. Salt beef and pork were prepared in Philadelphia and sent on to Trenton. Flour was also shipped from Pennsylvania until it became scarce. Attempts were then made to obtain flour from Maryland, but shipments were stopped by the British blockade of Chesapeake Bay. Quartermaster Greene also chartered a ship to bring rice from Charleston, South Carolina, but this also failed after British forces began blockading that port.[12]

Lack of forage caused cattle to become lean and unfit for slaughter. Much meat was spoiled due to transportation difficulties. As the weather grew warmer supplies of salted beef in camp became tainted. In April, rotten beef caused widespread sickness.[13] Difficulties with preservation and distribution caused hunger during most of the encampment. Even with local sources, there was never enough food to comfortably sustain the army at Middlebrook.

Show Me the Money

During the first half of 1778 the value of Continental money had been fairly stable at a ratio of about four-to-one with gold. By December, at Middlebrook, this ratio was down to nine-to-one and by April 1779 it had reached twenty-to-one. After that currency continued to decline at an accelerated rate until it reached an inconceivable ratio of about 1,000 to one in 1781.

This wild inflation rate had disastrous effects on General Greene's efforts to keep the army supplied. Farmers and merchants refused to sell their goods for currency that could lose half its value in a week. They preferred to barter, but the army had nothing of value to trade.

During the summer of 1778, before the army moved to Middlebrook, Quartermaster Greene had no credit and his rapidly-depreciating money was almost unacceptable. Only by paying up front in cash was he able to buy horses, wagons and forage, or hire civilians for any purpose. The few wagoners willing to serve also insisted on advance payment of their wages, and the farmers who might consider hiring or selling their wagons and horses to the army demanded immediate cash.

When the army began arriving in the area in the fall of 1778, Congress had virtually stopped using its paper money. As an alternative, it began allowing the individual states to provide supplies directly rather than make monetary payments. This system prevented a complete breakdown in the flow of supplies, but still did not provide adequate stores to sustain the camp. States were often late in filling their quotas; others contributed nothing. On January 21, 1779, John Pierce, the Paymaster General of the Army, reported to Washington that he was desperate for funds. By April, he stated that the treasury at Middlebrook was "entirely empty," and commented that "money does not pass."[14]

The decline of the Continental dollar caused inflation to skyrocket during the winter at Middlebrook. The Quartermaster and Commissary Departments had expenses of $5.4 million in 1776. This amount doubled the following year to over $9.2 million. By 1778 it had quadrupled to over $37 million. In May 1779, Congress's Committee on the Treasury estimated that at least $200 million would have to be spent that year by the two departments to continue to support the army unless finances could be put on a firmer basis. The core problem was simply the rapid depreciation of Continental currency.

As the value of Continental money dropped, Congress' solution was to print more of it, which drove the value down even further. By 1781 the exchange rate was $225 in paper money to $1 of hard currency. The average Continental Army private made $5 a month in Continental script and was seldom paid. Joseph Plumb Martin, a sergeant in the Connecticut line, remembered that after not having been paid for several months he tried to earn extra income by helping capture runaway slaves. He lamented, "The fortune I acquired was small, only one dollar; I received what was then called its equivalent, in paper money, if money it might be called, it amounted to twelve hundred [Continental] dollars, all of which I afterwards paid for one single quart of rum; to such a miserable state had all paper stuff, called-money depreciated."[15] Robert Hooper, Quartermaster General for the Pennsylvania Regiment, reported to General Greene at Middlebrook in February 1779, "Wheat is now from 12 to 15 dollars and rye in the same proportion and so great is the dislike of the [Continental] money that 7 dollars hard currency will purchase a bushel of wheat when 15 Continental Dollars are refused."[16]

What was most discouraging was that suppliers had the abundant provisions that the distressed army desperately needed, but would not accept the devalued paper money. Merchants and farmers had to survive and feed families, so they reluctantly sold their goods to the British Army, which could pay with hard currency.

During the months at Middlebrook many of the states preferred to direct food and supplies to their own militias, rather than supporting the Continental "national" army at Middlebook.[17] Washington's officers were often forced to compete with state recruiters who could offer recruits higher pay and additional land bounties. Congress had a very difficult time actually paying its soldiers at all. When the Continental Army called up militia units, they had to provide supplies for them out of their depleted stocks. At times, the tormented General Washington was forced to send citizen-soldiers back to their home states because they could not be fed or clothed from his army's dwindling supplies.

Keeping the Horses Alive

Forage or fodder, mostly hay and straw, was to the Revolutionary Army what oil is to a twentieth-century army. Army Historian Erna Risch agreed that "the heart of the transportation problem was forage supply system. Without the horses and oxen, the army could not move supplies or artillery, and without cavalry they would be helpless against onslaughts of the British Dragoons."[18] Lord Stirling, in command of the camp during Washington's absence in December 1778, ordered forage to be seized from local farmers. He found enough in Monmouth County to ease the emergency. But when he attempted to purchase it, local farmers refused to sell or were outraged by the meager prices offered by the army. [19]

By May 1779 the starvation of horses due to the shortage of forage caused transportation to come to a standstill. James Abeel, Deputy Quartermaster General, reported an example of the difficulty: "Several Waggons and pieces of artillery are now in this town

[Morristown]. I have not a pound of hay or anything else to feed their horse which are already so reduced that they can scarcely go. You can imagine how the service suffers at this place for want of forage."[20] Owen Biddle, Deputy Commissary General of Forage, attempted to rent pasturage from Somerset County farmers but many of them refused. As a consequence horses began to perish.[21] During May new horses began arriving at Middlebrook in preparation for the Sullivan-Clinton campaign against the Iroquois Nations, but they were soon requisitioned for duty in the camp.

Clothing, Tattered and Mismatched

Congress made a strong appeal for clothing materials from each state as soon as the troops began arriving at Middlebrook in the fall of 1778. Patrick Henry sent nine wagons full of uniform material to the Virginia forces. Supplies also arrived in Boston and other free ports from France. Most of these imports were finished goods, and included shirts, stockings, shoes, and blankets. These supplies of clothing, issued before going into winter camp at Middlebrook, soon wore out.

Distinctive uniform colors for each state did not exist until the Middlebrook Encampment. Before that time the most common regimental colors were green and brown. The familiar dark blue eventually became the standard of the camp. An early attempt to cloth everyone in functional fringed hunting shirts failed when not enough linen could be procured.[22]

The Clothier Generals Depot was located in Bedminster, New Jersey. The area today, adjacent to Pluckemin, is known as Burnt Mills. Most units at Middlebrook were dependent on supplies of clothing from their home state. Some states, notably Massachusetts and Virginia, were better providers for their regiments than others. The more affluent states sent abundant clothing supplies. For instance, in December 1778 three Massachusetts battalions received twelve wagonloads of materials imported through Boston, and paid for it in cash. Ragged troops from poor states were often purposely hidden in obscure campsites during visits by officials or foreign dignitaries.

Without a centralized system for procuring uniforms and other supplies, deliveries from states proved unreliable. New York troops suffered without shirts until March. When the apparel did arrive it was infected with lice and could not distributed.[23] After additional supplies, including shoes, finally reached the camp in April, the men were warned that lack of clothing was not an excuse for refusing to work.

Wool was scarce, but linen used in uniforms was handwoven in most colonial homes. Blankets often substituted for uniforms but were always in short supply. Apparel made of leather, such as breeches, jackets, and shoes, was more readily available. Overalls, which provided more durable protection to the legs, were preferred by soldiers for uniform trousers over breeches and stockings.

Supplying the army with shoes was a problem throughout the war. In 1776 Sergeant John Smith, on the retreat through New Jersey from New York City, reported in his diary, "our soldiers had no shoes to wair; was obliged to lace on their feet the hide of cattle we had kill'd the day before."[24] While clothing imports from France helped greatly at Middlebrook, procuring shoes remained a serious predicament. The quality of French footwear received was shoddy. The Board of War correspondence noted on December 6, 1779, "putting in small scraps and parings of leather and giving the shoes the appearance of strength and substance, while the soals were worth nothing and would not last more than a day or two's march."

Cattle slaughtered to feed the army provided an abundant source of leather for shoes. The Commissary of Hides, managed by Colonel John Meheim, Quartermaster General of the New Jersey Militia, was located in Bound Brook. This department collected, tanned, and finished the hides into leather suitable for shoes. The slaughterhouse was located in a building along the road opposite General Greene's quarters at the Van Veghten House.[25]

While hides were plentiful, shoemaking was still done by hand, and required skill. There were too few shoemakers in the Raritan Valley and Newark to produce the quantities needed by the camp. As the food system changed to salted beef and pork, provided from Morristown and Trenton, the ready availability of hides at Middlebrook diminished. The army never really solved the shoe problem, and it caused great suffering at all the winter encampments.

The Tragic Lack of a Centralized Supply System

May fresh horses began arriving at Middlebrook in preparation for the Sullivan-Clinton campaign against the Iroquois Nations, but they were soon requisitioned for duty in the camp. The efforts to deliver supplies to the Middlebrook encampment were always disorganized and unsystematic. The Continental Congress's attempts, hindered by the lack of any established system to equip and feed its army, was a failure. The states had the primary responsibility to provide supplies to their regiments, but these local attempts were also disorderly and pervaded by corruption. Continental and state purchasing agents competed with each other for resources. Prize cargoes from heavily laden British merchant ships captured by privateers along the New Jersey coast could have provided much of the needed supplies for the men at Middlebrook, but there was no system in place to haul these goods the short distance to the camp

Despite Nathanael Greene's endeavors, the lack of control of the supply system was the most tragic aspect of the second Middlebrook Encampment, as well as at other large scale winter camps at Valley Forge and Jockey Hollow. The Continental Congress, confronted with such an immense task, lacked the experience and funding to administer the military activities of the developing nation.

Chapter Seventeen

General Greene's Worst Nightmare

Supply Snags at the Camp

MIDDLEBROOK AND ALL other winter encampments of the American Revolution were plagued by the same chronic problems: insufficient supplies due to lack of funding and inadequate transportation. Underlying these difficulties was the apathy of a Congress lacking the power to tax. Early support for the war by a prosperous and motivated public began waning by the end of the first year of the war, the last months of 1775. The nation, with all its abundant resources, began complacently watching its army starve, and in fact often supplied the enemy if a greater profit could be made. The enduring indifference of civilians was fueled by the devalued Continental currency, coupled with rampant inflation. At Middlebrook shortages of food, forage, and clothing were amplified by the additional burden of equipping and suppling manpower for the upcoming campaign against the Iroquois Nations.

Major General Nathanael Greene had the daunting responsibility of feeding and supplying the American army at the camp during the winter of 1778-1779. Greene overcame incredible obstacles to achieve the most successful winter encampment of the American Revolution. He was the true champion of Middlebrook. This astonishing feat has been overshadowed by his achievements on the battlefield.

Greene was born into a prosperous Rhode Island Quaker family. He commanded the Rhode Island Militia when the war first began. After the opening battles of Lexington and Concord, in April 1775,

Nathanael Greene. *(Mezzotint full length by V. Green from a painting by Charles Willson Peale.)*

he was appointed Brigadier General in the newly-established Continental Army. He served under General Washington in all the early campaigns at Boston, New York, New Jersey, and the defense of Philadelphia. He was appointed Quartermaster General of the Continental Army in 1778. After Middlebrook, General Washington appointed Greene as the commander of the Continental Army in the southern theater. Washington had great respect for him and they became personal friends. Greene emerged from the war with a reputation as General Washington's most talented and dependable officer and is best known for his successful command in the southern states. Greene and his wife "Kitty" were favorite social acquaintances of George and Martha. After the war, he became a successful planter in the South. He died in Georgia in 1786 from a heart attack at age forty-three.

A Reluctant Quartermaster

During the previous winter of 1777-1778 at Valley Forge, Washington appealed to Congress to send members of the Board of War to witness for themselves the horrifying conditions. The officials were appalled when they observed many soldiers dying from starvation, or from exposure to the cold and sickness. They frantically reported back to Congress that "the Appointment of a Quarter Master General is a matter of great Importance and immediate necessity."

General Washington encouraged General Nathanael Greene, one of his most valued field commanders, to accept the new post. Greene at first refused the job. As owner of a small business before the war, he knew of the difficulties that the job entailed and believed that the position was beyond his ability. He also loathed being transferred to a staff post from a line command where he led men in combat. He wrote to Washington of his concerns: "The little support we have had from Congress, the difficulty of procuring aid from the different States renders the business very difficult to manage. These embarrassments together with the bad state of our money, and the difficulty of procuring assistance from the line of the army has discouraged me. In this employment I have very little to gain and everything to lose."[1]

Greene remained adamant in refusing to relinquish his revered position alongside Washington in battle. Washington persisted, telling him that in his opinion he was the only officer who could get the job done. Greene grumbled to fellow General Alexander McDougall, "All of you will be immortalized in the golden pages of History, while I am confined to a series of drudgery to pave the way for it."[2] The dedicated Patriot finally recognized that the post was critical to the raw army's survival, and reluctantly accepted the demanding job.

When the army moved into Middlebrook for the winter in late 1778, it was immediately confronted by alarming supply problems. Food, supplies, horses, wagons, and forage were needed quickly, since it was apparent to Greene that the army would not survive the coming winter season. He was aghast when he learned that he had to rely on greedy purchasing agents who inflated their commissions. Forage, food for the horses, which was the heart of the all transportation, remained especially scarce. The army at Middlebrook moved by horsepower and horses had to be fed.

Greene takes Charge Amid Adversity

Nathanael Greene assumed his new duties on March 23, 1778. His first challenge was to solve the lack of land transportation. Carts, wagons, and sleighs, as well as the men who ran them—wagoners or teamsters—were in short supply The situation had been so dire at Valley Forge that men had constructed small carriages and yoked themselves to them, or had carried stacks of logs and provisions on their backs. Greene threw himself vigorously at the problem. Fresh supplies were critical to survival for the rest of the winter, and spring, with its increased demand for supplies for a new campaign, would be coming soon.

Greene wisely decided to live among the troops at Middlebrook, rather than manage remotely from Philadelphia. Living amid the hectic day to day provisioning activities allowed him to assess first-hand the myriad bottlenecks in the supply chain. He moved into the Van Veghten House in Finderne, adjacent to the quartermaster storage depot.

He soon concluded that more management and manpower were needed to procure and move supplies. He hired new purchasing agents that he hoped would be more trustworthy. This sudden growth in the number of staff personnel caused a significant increase in costs. By 1778, only six weeks after the army arrived, supply expenses at Middlebrook had already quadrupled to more than $37,000. This was the equivalent of roughly $700,000 today.

Before moving to Middlebrook, Greene drew up new procedures for solving the vexing problem of procuring wagons and teams, and for enlisting wagoners. He always considered himself as working directly for George Washington and the army rather than following directions from Congress. It was evident to him and other military leaders that the legislators in Philadelphia had limited knowledge about how an army operated, or what was required for a long-term war effort. Congress naively created a compensation plan for Greene that was designed to incentivize him. Oddly, his salary was based on commissions on the amount of supplies he purchased. However, when supply expenses rose exponentially, due mainly to the hiring of more staff, Congress grew suspicious. Early in 1778 Greene was falsely accused of collecting exorbanent fees by artificially inflating purchases.

The infuriated general denied these allegations and threatened to resign. After two Congressional investigating committees found him innocent, and General Washington had swore to his competence, honesty, and effectiveness, he was exonerated and persuaded to stay on. Greene's recruiting program for wagoners eventually was successful, and was a major reason the army was able to endure at Middlebrook until spring.

Washington had words of praise for Greene when he reported to Congress: "by his conduct and industry [the Quartermaster Department] has undergone a very happy change, and such as enabling us, with great facility, to make a sudden move with the whole Army and baggage from Valley Forge in pursuit of the Enemy [to Monmouth, Courthouse] and to perform a march to this place."[3]

Strategically Ideal but Challenging to Supply

While Middlebrook's safe location nestled in the Watchung Mountains was ideal for defense, it was an arduous task to keep the remote site supplied. Two depots—at Trenton and Morristown, a considerable distance from the camp—provided supplies. The "magazine" at Trenton was twenty-five miles away, and the quartermaster store at Morristown was eighteen miles distant. These depots furnished camping equipment, clothing, food, and forage to Middlebrook. The storage centers appear to have duplicated each other. Both had almost all of the materials the army used.

There was a critical shortage of wagons in Trenton and Morristown to haul stockpiled stores to Middlebrook. The low wages that the army offered attracted inexperienced teamsters. They were signed on for only a year, not enough time to train them adequately.

In January 1779, the number of wagoners to supply the entire camp had dwindled down to only to sixty. The delays caused by the lack of means to deliver necessities were so desperate that soldiers were drafted to replace the civilians to drive wagons. In May, sixty wagons of food destined for Middlebrook were loaded at Trenton, but no teamsters were available to deliver the perishables and the food had to be discarded.[4]

During the first months of the encampment the bad weather paralyzed the movement of provisions along the narrow dirt roads. With all supplies transshipped to Middlebrook from the distant centers, rough country roads and the capricious New Jersey climate determined the flow of all essentials. The mud paths leading from the Middlebrook camp were in deplorable condition. By January most were impassible. The roads to Trenton were extremely bad and had deep ruts worn in them caused by constant use. The ice and mud limited shipments from Morristown but that road was reported to be in better condition. Fortunately, the vital road covering the seven miles from the main camp to Pluckemin, now Route 202, was reported to be in good condition.

General Wayne blamed lax maintenance for the problem. He wrote to militia lieutenant John Van Nest on January 16, 1779, "The state of the publick roads in the vicinity of this camp is known to be extremely bad Thro' the neglect of the supervisors or officers whose duty it is to keep them in proper repair. It is with the utmost difficulty provisions and other necessary can be carried through to the use of the Army."[5]

A heavy snowstorm and low temperatures set in on December 23. This was followed by another snowfall on January 5.[6] Farmers in the area were enraged when soldiers began stealing their fence rails for kindling. Carpenters at the Morristown depot tried building sleds in a desperate effort to get supplies through to Middlebrook. The sleds proved to be so successful that several more were built, and they became the main means of transport to the camp for several weeks. By the end of January, the weather improved and remained fair for the remainder of the encampment. Surgeon Thacher remarked in his journal, "We have passed a winter remarkable mild and moderate; Since the 10th of January we have scarcely had a fall of snow or a frost and no severe weather. At the beginning of this month [April] the weather was so mild that vegetation began to appear; the fruit trees were budded and stand in full blossom."[7]

Corruption in the transportation system also served to hinder supplies from reaching the encampment. Stores arriving at Morristown for transshipment to Middlebrook were often opened and pilfered in transit. It was common to have barrels of flour arriving only half full. The badly needed flour was being stolen by wagoners and sold to local inhabitants on the way to Middlebrook. In May, nine civilian employees of the army were found guilty of stealing from the wagons.

Dishonesty among civilian workers and soldiers was driven by their pay not being adjusted for the wild inflation. This issue depressed morale and grievances quickly mounted. General Arthur St. Clair the commander of the Pennsylvania Line, warned Joseph Reed, President of the Pennsylvania Council, of the dire situation, "...judge from the rest of this circumstance, a dozen eggs cannot be purchased for less than two dollars."[8]

In January a new issue of currency was provided to the soldiers at Middlebrook in exchange for their devalued money. The British, through their Loyalist informants, were kept aware of the depreciation problem at the camp. Counterfeiters began flooding the area with the old issue to be exchanged for the new in an attempt to bankrupt the army.[9]

After having struggled to keep the army fed and clothed throughout the winter, Nathanael Greene soon saw an opportunity to return to his former combat role. In August 1779, only one month after the army left Middlebrook, he submitted his resignation as Quartermaster General. He left knowing that Washington believed he had done an outstanding job, and that this had enhanced his reputation throughout the army and in Congress.

Washington gave Greene a letter of commendation after the encampment at Middlebrook:

When you were prevailed on to undertake the Office in March 1778 it was in great disorder and confusion and by extraordinary exertions You so arranged it, as to enable the Army to take the Field the moment it was necessary, and to move with rapidity after the Enemy when they left Philadelphia. From that period to the present time, your exertions have been equally great, have appeared to me to be the result of system and to have been well calculated to promote the interest and honor of your Country. And in fine, I cannot but add, that the States have had with you, in my opinion, an able, upright and diligent Servant.[10]

He was soon appointed as Commander of the Southern Army and was recognized as second only to Washington among the field commanders of the Continental Amy. Greene emerged from the war with a reputation as General George Washington's most gifted and dependable officer. His fame as a combat officer and his successful command in the southern theater is revered, but his contribution to the cause of American independence as Quartermaster General during the Middlebrook Encampment, a critical period of the war, has been neglected in history.

Chapter Eighteen

Civilian Support

The Patriots of the Raritan and Washington Valleys

THE PRESENCE OF supportive inhabitants in the vicinity of large winter encampments was a critical factor when selecting camp sites. The almost 10,000 troops that suddenly arrived in Somerset County in late 1778 were never self-sustaining. The Middlebrook Encampment was dependent on the citizens of Somerset County for grain, forage, and other supplies, but local resources were limited in the sparsely inhabited countryside. There were as many soldiers in the camp as there were in the entire county.

Typically, soldiers at the winter encampments were bitterly remembered by their civilian neighbors for acts of fence burning, pillaging and drunkenness. But Middlebrook differed from all other major winter encampments of the war. Fortunately, there was little friction and hostility with local people, and a remarkable spirit of cooperation existed between the army and nearby residents. This was true at the main camp and at the neighboring artillery post at Pluckemin, despite the fact that these civilians had been ensnared for years in the depressing conflict and were in need of the same basic necessities that the army required.

In fact, Somerset County residents had much to gain from the army's presence. The large armed force provided them with security and protection. Their land that had been ravaged by enemy raids during the forage war of early 1777, and during those six months their homes were plundered and burned, and their patriotism tested. When Middlebrook was created in December 1778, General

"Colonial Blacksmiths Forging Muskets for the Minutemen at The Outbreak of the American Revolutionary War" (19th century wood engraving)

Washington notified New Jersey's Governor Livingston that the safety of his people was of "particular consideration."

For the first time there was an opportunity for alternative employment in adjacent farming communities. More than fifty civilians from nearby towns were hired during the encampment to cart hay, drive wagons, and tend and slaughter livestock. Area residents were assigned to important jobs. Jacob Weiss managed the quartermaster stores. John Melheim was the Commissary of Hides, which were used to produce leather goods, and Cornelius Voorheis was a bookkeeper who paid civilians for their services.[1]

The army's presence also created an active and profitable market for surplus produce that could be sold at inflated prices. Housing was scarce in the area, so families rented or shared their homes with officers. The army leased buildings from residents for barracks and store houses, and pastures for horses and livestock. With few exceptions living and working together fostered enhanced an agreeable relationship with the local residents.

The loyalty of the people of the Raritan and Washington Valleys went beyond simply being accommodating. All males between the ages of sixteen and sixty formed the effective New Jersey Militia regiments. With modest support from the Continental troops sequestered in Morristown, these citizen-soldiers had repulsed the Crown forces during the six month of forage war of 1777. They not only defended their homes but went on the offense by relentlessly harassing the British to prevent them from supplying their huge army and the thousands of Loyalists stationed (living) in New Brunswick. The Redcoat invaders required an enormous quantity of provisions and were forced to live off the land. The failure of their foraging, due in large part to the resistance of the New Jersey Militia, was the main reason that they eventual abandoned New Jersey. These Patriots were the civilian neighbors at Middlebrook.

Arduous Civilian Life

The stillness of the Washington and Raritan Valleys was only broken by the lowing of cows, an occasional musket blast somewhere in the camp or the clip-clop of a horse's hooves. Life was hard and pleasures were few for the residents of the Somerset County towns near the encampment. Families grew all their own food. In particular, a great deal of corn was consumed. It was roasted, boiled, and baked into cornmeal bread. Hearty vegetables, such as squash and beans, were also on the dinner table. The farm families fashioned their own clothes from flax and wool, so almost every home had a spinning wheel and a loom. Women sewed and knitted constantly. The ill-fitting clothing they produced itched and chafed, and was too hot in summer and not warm enough in winter.

The air in the valleys was often filled with wood smoke from cabin chimneys. Lime was tossed into outhouse pits in an attempt to reduce the stench. Many sicknesses meant long suffering or even death. People relied on folk medicine and Indian cures such as herbs, tea, honey, bark and roots. Overall life expectancy was about forty years. Women typically bore between five and ten children in their lifetimes. Burdened with making clothing, gardening, preserving

On the home front. *(Library of Congress Collecion)*

food, and other tiring and endless chores, they aged rapidly and died young. Infant mortality was so high that to avoid any emotional attachment parents typically did not name their offspring until the child reached two years of age.

Men and boys hunted for rabbits, squirrels, bears, and deer in the Watchung hills. They fished in the Middle Brook and the Millstone and Raritan Rivers. Most boys and girls worked hard alongside adults. Their lives, like their parents, were tied to the seasons. They plowed the fields, mended fences, skinned animals, fished and hunted. Adults avoided the stagnant water sources but managed to slurp down four gallons of corn whiskey each year, supplemented by cider and beer. Children also imbibed.

Occasional Clashes with Civilian Neighbors

Army leaders who had served during the previous winters at Morristown and Valley Forge, where they met with hostility from nearby residents, anticipated resentment and hostility at Middlebrook. General Washington issued a warning two weeks after arriving there: "The strictest regularity and decency is to be observed [toward the people] of the country and officers are desired to use their utmost endeavors to detect and bring punishment to marauders. He also warned nearby citizens to avoid hunting in the neighborhood of the camp as this might alarm the soldiers.[2] His first order when he returned from Philadelphia was to request the troops to respect the property of the farmers in the vicinity of the camp.

For the most part, relations with civilians in the Middlebrook area proved to be agreeable, but there were a few unfortunate incidents. The Pennsylvania Regiment, quartered across the Raritan River opposite Millstone, arrived late. They had insufficient fuel to weather the severe cold spell at the end of the year, so they tore down fences to use as firewood. Lack of firewood was a vexing problem in all the winter camps during the entire war. Washington admonished the men for this offence by issuing a General Order:

Head Quarters, Middlebrook, December 29, 1778

The many positive orders relative to the preservation of inhabitants fences render it painful for the Commander in Chief to repeat them; but the frequent complaints which are daily exhibited to him of the wanton destruction of the enclosure by the soldiers compels him to urge officers of all ranks to search out and bring to immediate and severe punishment every soldier who presumes to burn or otherwise destroy rails or any part of the farmer's enclosures. Honor and humanity dictate that we should carefully preserve the property of our fellow citizens.

Fence destruction continued despite this warning. General Anthony Wayne, commander of the Pennsylvania Regiment, added his own edict when the situation worsened:"Every possible method will be employed to stop the destruction." He chided his men by reminding them that other regiments at Middlebrook had left fences undisturbed in their assigned campsites.[3]

The trampling of wheat and rye fields by soldiers moving into the camp area especially angered farmers. Most of the grievances originated from the Maryland Regiment, whose camp was adjacent to Bound Brook and extended west along what is now Route 22. This offence occurred so often that in February it was necessary to post guards in each field. Citizens also complained that the Maryland camp was filthy.[4]

The affable General Greene appreciated the value of creating a favorable impression with the local population. He wrote to General Wayne in December 1778: "One thing is to be said of the inhabitants in this part of the country, they are generally good whigs [Patriots] and deserve all the indulgence that can be given them. Besides which it will be in our interest to keep them as good natured as possible for our own case and convenience."

Unfortunately, his amicable relations with the public soured when he was besieged by several petty lawsuits in January 1779. These complaints were filed by local justices of the peace seeking damages on behalf of their residents for the seizure of grain by soldiers. By April, Greene was so embittered that he described them as "men who show their learning and improve their genius by swarming about us like birds of prey seeking whom they may devour."[5]

Greene's relationship with local officials reached a low point in March. Abraham Van Nest, a justice of the peace, attempted to arrest two soldiers of the Pennsylvania line who were accused of stealing chickens. Van Nest stormed up to Greene's house late at night, banged on his door, and demanded that the soldiers be arrested. The enraged Greene responded by calling Van Nest a damned rascal. He then had him escorted back to his home under guard.[6]

Famished soldiers seizing food from local farmers caused the most enmity. American foraging parties reluctantly raided the homes of their countrymen and stole grain and livestock from destitute families. This "friendly foraging" so antagonized residents that the contrite army offered to reimburse farmers by giving them their own requisitions for the pilfered items. Farmers retaliated by grossly inflating values when they redeemed the invoices.

Problems with civilians were always detected by Loyalist spies living near the camp. During that time period a series of American defeats in the South was depressing the nation's morale. An elated New Jersey Loyalist source reported, "The inhabitants feel the effects of the war in general want of provisions of all kinds. The depreciation of paper dollars is a blessed thing: it does the work as well as a drubbing would, but a little of both would be most effectual." Another Tory spy gleefully reported, "Governor Livingston quarrels with all the military in Jersey. They must know the game is up, the people will no longer be abused, robbed, deprived of every enjoyment of life to serve the ends of a few."[7]

Civilians also profited by the insatiable demand for liquor in the camp. Illegal spirits were openly traded and smuggled to the soldiers. General Washington protested to Governor Livingston in May 1779: "I am ever now and then embarrassed by disputes among the officers and inhabitants, which generally originate from the latter coming into camp with liquor and selling it to the soldiers and as the officers allege, taking clothing provision or accoutrements in pay."[8] General Wayne was finally sent to the New Jersey Legislature to plead for a law prohibiting the practice. It was never enacted. A search of the homes of civilian merchants in Morristown turned up clothes, tents and other articles taken from the needy army.

The most flagrant act of civilian defiance at Middlebrook occurred in May 1779. The army urgently needed three empty barns near Millstone to hospitalize influenza victims. The barns were not in use, so no difficulty was expected in commandeering them. The army's request was rudely rejected by Earnestus Harlingen, justice of the peace for the town. He threatened to imprison the first person "who shall prostitute a barn for the use of the sick soldiers." Armed guards seized the barns, rather than allowing the sick soldiers to lie in open fields. The reason for Harlingen's hostility is unknown.

Despite these few unfortunate incidents, relations with civilians in the Middlebrook area were mostly agreeable. As General Washington was preparing to leave, he received a heart-warming letter from the ministers and members of the Reformed Church in Raritan that reaffirmed their support for the cause of independence and their respect for the army. Washington was delighted with the letter and replied, "In quartering any army and supply its wants, distress and inconvenience will occur to the citizen, I feel myself happy in a consciousness that these have been strictly limited by necessity, and in your opinion to my attention to the rights of my fellow citizens. I thank you sincerely for the sense you entertain of the conduct of the army."[9] When the Continental Army departed it graciously left behind instructions to compensate residents and to settle claims for any material that had been requisitioned during the encampment.

Chapter Nineteen

The Army Departs

But the Tumult Goes On
Middlebrook, 1779-1781

DURING THE SECOND Middlebrook Encampment, the occupation from December 1778 through June 1779, the number of troops quartered on the site reached as high as 10,000. They were spread throughout the Raritan Valley and the flatlands below the first Watchung ridge extending east to the Raritan River. General Washington believed that this area was secure and that there was less need to be vigilant than in 1777. The British Army had remained sequestered in New York City for the past year and a half. There was little probability of an invasion of New Jersey to advance down the New York-Philadelphia corridor. The Raritan Valley had also offered better roads, a civilian workforce and more abundant resources to Washington and his troops.

The Washington Valley, behind the first ridge, had been the center of activity during the First Middlebrook Encampment the previous year, in the late spring and summer of 1777. The Middle Brook courses the length of the valley, encompassing the towns of Stirling, Warren Township, Millington, Liberty Corner and Pluckemin. This strategic area saw little military occupancy during the second stay at the Middlebrook encampment. However, it remained crucial to the defense of the main camp, and was ready to protect the interior of the state if an attack did occur. Historians have overlooked the strategic role of the Washington Valley after the 1777 encampment.

In 1912, the *Somerset County Historical Quarterly* reported on the fortifications in the Washington Valley around the Middlebrook camp:

> Three strong earth forts were thrown up guarding the entire valley. One is near the Middle Brook which is still in a good state of preservation on the farm lately owned by Mr. Kennedy Bolmer. The second fort was in the center of the valley, and the third on the road from Martinsville to Pluckemin. Both of these forts are now entirely destroyed. A fort was also built on the hill opposite Chimney Rock to guard the road leading through the narrow mountain gorge winding along the Middle Brook. Cannon were also planted on the hill summit looking toward the valley of the Raritan.[1]

The earthen redoubt near the Bolmer Farm protected the rear and right flank of the Middlebrook Encampment. This strategically placed earthworks fortification still exists in Martinsville. Its cannons, trained down on the pass at Chimney Rock Road, served as the principal defense position between the southern end of the ridges and Springfield, twenty miles north. It was constructed during the First encampment and is described in Chapter 5.

After the encampments the crest of the first ridge, with its lookout posts on the rocky promontories, continued to serve as the vantage points over the central New Jersey plains. The ridgeline was also used to support a system of twenty-three signal towers used to warn American posts of any hostile activity in Somerset, northern Monmouth, Essex and Bergen counties, as well as in the Hudson Valley. They could be seen from the coastal highlands of Monmouth County all the way to Elizabeth. Two of the beacons was situated on the campground on high points on the ridge at Vossellor Avenue and the Chimney Rock gap.

As mentioned, the winter in central New Jersey during the second encampment had been mild. Snow did fall in late December and early January, but after that time it was generally above freezing. It warmed in the spring a couple of weeks earlier than usual, with fruit trees flowering in early April.

French and Spanish ambassadors visited the encampment in March and pledged support from their respective countries. A grand review ceremony was held in their honor. Baron Von Steuben continued to develop the American army into a disciplined, combat-ready force. In the spring General Washington had ample time to plan and gather supplies for the next campaign. The most significant progress made during the Middlebrook Encampment of 1778-1779 was that the American Army matured into a professional fighting force whose proficiency matched the other great armies of the world.

Final Days

The camp began breaking up in May. General John Sullivan's Brigade of 4,000 troops was the first to leave, heading north on their campaign against the Iroquois Nations in western New York State. The Pennsylvania and Virginia regiments headed out for Fort Pitt on May 6 to join Sullivan on the New York campaign. Later that month, additional troops were ordered away to new posts and other campaigns. The New Jersey troops under General Maxwell proceeded to Easton, Pennsylvania to join Sullivan.

General Anthony Wayne, commanding the Pennsylvania Line on the south side of the Raritan River, also broke camp in the last days of May. His regiment marched to Stony Point in the Hudson Highlands. His Pennsylvanians surprised and captured the British bastion there to achieve a brilliant victory on July 15, 1779. The Virginia Line also headed north to the Hudson Highlands. The total evacuation of the Middlebrook Encampment was completed by June 3, 1779. All huts were abandoned.[2]

The final weeks at Middlebrook were difficult. Recruiting efforts remained unsuccessful and continental currency continued to plummet in value. The desertion rate was appalling, and shortages of food and clothing continued. The pleasant weather and social events of the spring did not improve General Washington's gloomy mood as the army prepared to leave. He wrote on May 8, "our army as it now stands is little more than the skeleton of an army and I hear no steps being taken to give it strength and substance."

Washington also knew that when the army left Middlebrook the entire state of New Jersey would not be protected. He warned Governor Livingston that the state militia was now his only means of defense: "The Militia must...be assembled in time to prevent any inconvenience from the departure of the troops"[3]

As the remainder of the Continental Army headed to Morristown, on the way to West Point, Washington feared that his debilitated columns might be attacked along the way. A British spy reported on June 3 that "Mr. Washington's Army marches fast and by such different routes. I cannot ascertain his numbers but cannot make them above 5,500." This report shows that the enemy was completely baffled by the hasty and well-ordered departure of the Americans from Middlebrook. At least 10,000 troops marched from the camp between June 1 and June 3. An attack on the strung-out columns never did occur. The king's army was then engrossed with bringing the war to the southern states, a sweeping change in their strategy.

General Washington's personal departure from Middlebrook on June 3 is noted in his letter to the President of Congress: "We shall press forward with all diligence, and do everything in our power to disappoint the enemy. I expect to set out this day towards the Highlands by way of Morristown."[4]

The deserted huts and other structures at Middlebrook were soon dismantled by the local people and used as construction material or for firewood. A few buildings continued to be used as a hospital during the winter of 1779-1780. According to local archeologist A.A. Boom, the remains of some stone defenses could still be found in the 1970s in places where they had not been disturbed by housing developments. However, many of these ruins later proved to be foundations of civilian structures.[5]

Simcoe's Audacious Raid

In October 1779, five months after the Continental Army left Middlebrook, a significant military event occurred in the Raritan Valley. Loyalist spies reported to British General Henry Clinton that the Continental Army was preparing to invade New York City, the headquarters of the Crown forces in America. It appeared that the plan was for an amphibious assault, with the Patriots crossing the Hudson River in flat-bottomed barges or bateau, the landing craft of that era.

The shocked Clinton received further intelligence that fifty bateau were being assembled on the Raritan River at the Van Veghten Bridge, where today the bridge on Finderne Avenue in Bound Brook crosses the Raritan River. Clinton swiftly responded to this threat by sending Colonel John Simcoe on a daring mission deep into enemy territory. Simcoe, one of Clinton's most competent young officers, was given the two objectives. He was to destroy the bateau to prevent the invasion, and was to capture New Jersey Governor William Livingston, who was staying nearby in Bound Brook.

Simcoe, a twenty-nine-year-old combat veteran, commanded the Queen's Rangers, an elite, highly trained infantry unit, renowned for their effective and ruthless surprise attacks. Earlier in the war Simcoe saw action during the Siege of Boston, the New York campaign and the New Jersey and Philadelphia operations. At the Battle of Brandywine in 1777, according to legend, he ordered his men not to fire on three fleeing rebel officers. George Washington was one of them.

Simcoe's Rangers had a record of barbarity. Their standing orders for the foray into New Jersey: "Go—spare no one—put all to death—give no quarter." Earlier, in 1778, on a foraging expedition in Salem, New Jersey they raided the home of Judge William Hancock, which housed the Patriot militia. Bayonets were used in the stealthful night attack. Ten militiamen were slaughtered in their sleep and five others were wounded. Later that year, in what is today the Bronx, New York, Simcoe led a massacre of forty Native Americans serving with the Continental Army.

At 3:00 a.m. on October 26, 1779, Simcoe's unit of 300 Queen's Rangers was ferried from Staten Island to South Amboy at the mouth of the Raritan River. Simcoe left a force scattered behind in the present-day towns of Sayreville and Franklin Township to cover his withdrawal, then rode off along the Raritan's south bank with eighty mounted men. Their green jackets and tan breeches resembled the uniforms worn by American Major Henry Lee's Continental dragoons. This American regiment was quartered nearby in Englishtown. Simcoe's rangers cleverly masqueraded as the American horsemen and were able pass unopposed through twenty miles of Patriot held territory. Simcoe himself posed as one of Lee's officers.

At the Frelinghuysen Tavern on Main Street in Bound Brook he enhanced the deception by accusing the residents of being Loyalists. The Queen's Rangers rode on to Quibbletown, where they dismounted in order to pursue men they believed were Patriot militia. The Rangers then rushed to Phillip Van Horne's house in Bound Brook where Governor Livingston was living, but found that he had escaped. Next, the raiders burned down the Dutch Meeting House in Finderne. Simcoe was eventually recognized. A shot, fired at him from across the Raritan, alerted the local militia that the dreaded Rangers were nearby.

When the British troopers arrived at the Van Veghten Bridge they were disappointed to find only eighteen bateau. They burned the boats using incendiary grenades, then rode eight miles south along the river to Somerset Courthouse, today's Millstone. There they freed loyalist prisoners jailed in the courthouse and afterward burned the building to the ground.

Simcoe, not finding any other targets of opportunity, set out to return to South Amboy. He asked a young boy, who admitted that he was not familiar with the route, to serve as a guide. They galloped east on what is today Hamilton Road on Route 514 through Franklin Township. They intended to turn toward New Brunswick at a crossroads at the Garrett-Voorhees farmhouse. But the house had been burned down early in 1777 during the forage wars and little

The Army Departs

evidence of it remained. This caused the Rangers to miss the left turn and they were soon helplessly lost.

Desperate Simcoe heard musket fire behind him and was ambushed by an American militia unit of thirty-five men led by Captains Moses Guest and William Mariner. His horse was shot and fell on him as he was captured. The ambush occurred two miles from New Brunswick, near De Mott's Tavern at the intersection of Demott Lane and Route 514, now the site of the Stage House Tavern. Today the exact place of the skirmish is on Battle Place in Franklin Township. Simcoe described the unfortunate event in his journal.[6]

Determining to pass through this opening, [in a fence] I saw some men concealed behind logs and bushes, between me and the opening I meant to pass through, and I heard the words 'now, now' and found myself, when I recovered my senses, prisoner with the enemy, my horse being killed with five bullets, and myself stunned by the violence of my fall.[7]

Captain Guest reported:

Soon after getting on the road leading from the Millstone village to the bridge, I was informed by an express that the enemy was within a few hundred yards of me. I had just time to get to an open piece of woods, when they made their appearance. We attacked them as they came up, but they came on so rapidly that we could only give them one discharge. Colonel Simcoe's horse received three balls, fell on him and bruised him badly. I left a physician with Simcoe, and proceeded on.[8]

A militiaman was about to bayonet Simcoe as he lay on the ground, barely conscious. Captain Mariner deflected the blade and exclaimed, "Let him alone, the rascal is dead enough." Two months later Simcoe was freed in a prisoner exchange and sent to Charleston, South Carolina. While there, he learned that William Mariner had been captured. The appreciative Simcoe wrote to General Clinton and arranged a parole for his savior.9 Simcoe later joined Cornwallis' army at Yorktown in 1781 where he again became a prisoner of war. After the war he retired to his estate in England. He was elected to Parliament in 1791 and was appointed Lieutenant Governor of Upper Canada that same year.

The Iconic March to Yorktown: Allied Armies Converge at Middlebrook

In the summer of 1781, two years after Simcoe's raid, Middlebrook would again come to life. At that time most of the allied army, a force of more than 16,000 American and French troops, had been stalemated for two years while blocking the British from leaving New York City. They were confronted by 14,000 Crown troops in the city. During their five years of occupation the British had turned the city into a well-fortified bastion. A brash decision was made By General Washington and the French commander Rochambeau to move the allied forces 500 miles south to Yorktown, Virginia rather than attack the heavily garrisoned city. They thought that it might be possible to trap half of the British Army at Yorktown, a peninsula near the mouth of Chesapeake Bay. Thousands of the Franco-American troops rushing south toward this destination, tramping through and briefly occupying the old camp at Middlebrook.

This happened in August 1781, during the seventh year of the war. By that time, the citizens of the towns in the Raritan and Washington Valleys surrounding the Middlebrook campsite had become more secure and optimistic. The British had become lethargic in New York and the threat of being plundered again by ravenous Redcoat hoards seemed remote. The French had now joined the American cause with their strong, well-trained army and massive fleet.

After crossing the northern border of New Jersey during the last week of August 1781, the Allied army divided into three columns and headed south. The infantry column, led by General Washington, passed through Newark, Elizabeth, Westfield, and South Plainfield on their way to Bound Brook. A second large and important column, carrying supplies and weapons, followed a more secure route behind the first ridge. On August 27, 1781 they hurried down Valley Road through the Borough of Watchung and Mountain Boulevard in Warren Township and turned east on King George Road to pass through the middle of the Middlebrook Encampment.

French Lauzun Legion (Hussars) Officer, full (parade) dress, 1781. Before departure from Newport to Virginia. *(Courtesy Revolutionay War Forum)*

The French Army made up the third column. It passed by Morristown and followed what are now Interstates 287 and 78, marching ten miles to Liberty Corner, Bernards Township. The gas station at Exit 36 off of Interstate 78 West, near Liberty Corner, stands on a French campsite. They camped overnight in Liberty Corner at Bullion's Tavern on the grounds of the "English Farm." An historical marker on Valley Road at this location reads "On this site French Troops under Le comte de Rochambeau encamped August 29, 1781, en route to meeting General George Washington and achieving their victory at Yorktown" It is one of few campsites in New Jersey that remains on undeveloped farmland.

The next day the French parade passed directly through the Middlebrook Campground to their next camp at Millstone. To get there they followed Newman's Lane, Steele Gap Road, and Foothill Road. When the French column reached the Van Veghten Bridge, on Finderne Avenue, in Bridgewater, it turned south through Bound Brook and Manville to reach Millstone. Today, historical markers can be found along this route.

For an entire week, thousands of men, horses, and oxen of the allied army swarmed through the main streets of little Somerset

County hamlets that typically had only a church, an inn, and a few houses. Farm families came from many miles around to line the dirt roads along the routes to watch the spectacular parade. Most of the residents had never been further than the next farm.

The farm families were fascinated by the superbly clad troops. They were accompanied by thunderous drums and brass bands and some spoke an exotic foreign language. The families cheered wildly and offered the soldiers freshly baked bread, jellies, cheese, cider and liqueur made from cherries. Flirtatious country lasses ran out into the lines to greet the marching men. The well-disciplined French soldiers wore immaculate white uniforms with blue, pink, or green facing. Officers, noblemen mounted on stately steeds, wore silver helmets and purple sashes. It was a stirring experience for the humble Patriots and an event they would boast about for the rest of their lives.

In sharp contrast, the haggard soldiers of the Continental Army came by with bare feet and clad in the remnants of tattered uniforms or ragged civilian clothes. The citizens were appalled by the appearance of these emaciated American soldiers. They ranged from children under fourteen years old to men in their sixties. They were described by everyone as being in good spirits, despite their appearance. Many had been on campaigns for over five years.

The logistics of providing thousands of men that passed through New Jersey that one week with food, forage, firewood, and shelter made it necessary for people in Somerset County to assist the army. The support of citizens in the towns of the Raritan and Washington Valleys and other New Jersey settlements was crucial to the success of the march to Yorktown.

Many enchanting accounts of the five-day march through New Jersey on the way to Yorktown come from French officers and soldiers. They viewed New Jersey as an exciting wonderland filled with surprises. American soldiers left few descriptions of the march. For them the terrain, towns, and people were familiar and were much like the places they lived and not worth describing in writing.[9]

All allied forces came together at Middlebrook to begin one of the most renowned troop movements in the history of the world. They continued their trek to Yorktown, and in the end won the final decisive battle of the Revolutionary War. After the victory at Yorktown the Allied armies returned north and camped again in the Middlebrook area. The French camps in 1782 were at York Road in Millstone. It was likely the same site they occupied on their way south to the battle in 1781.

Dr. Robert Selig, project historian to the National Park Service for the Washington-Rochambeau Revolutionary Route National Historic Trail Project, reported that the Americans stayed in Bound Brook along today's Foothill Road and Steeles Gap Road to the west of the Chimney Rock Gap.[10] The order book of the New York Brigade reads, "Continued our march to Rariton River near Bound Brook and Encamped at Steels Gap."[11] This was the site of Steeles Tavern in Bridgewater and was within the camp of the Virginia Regiments in 1778-1779. Selig believes there are "thousands of artifacts" to be unearthed in the area. It is believed that Steeles Tavern was located adjacent to a spring head which was a water source for Allied troops.

These epic marches through the Middllebrook Campground were the last military activity that would occur on the historic site. It would languish in obscurity for the next two centuries.

PART III

Exploring a Dispersed Campsite in the Raritan and Washington Valleys

Chapter Twenty

"Rank Hath Its Privileges"

Officers Compete for Scarce Housing

DUE TO A SHORTAGE of private homes in the immediate vicinity of the Middlebrook Encampment, most senior officers had to live a considerable distance away from the main camp. Somerset County was a sparsely inhabited area with few large comfortable homes, so the availability was limited. General Washington set up headquarters in Somerville, five miles west of the camp. When the army first arrived, most officers were billeted in private homes in nearby villages. Apparently, the patriotic homeowners supported the war, and compensation was provided to them. There is no evidence of any resistance to sharing homes with the army.

After the soldier's huts had been completed many lower ranked officers were ordered to report back to camp to live with their regiments. Only generals, majors, colonels, and field grade officers were permitted to live outside the camp in civilian homes. Since only the most affluent families had homes large enough to accommodate the commanders and their staffs, the officers resided with the most prominent families in Somerset County. On December 19, 1778 *The Royal Gazette*, a New York City Loyalist newspaper, reported: "By late accounts from New Jersey, we are informed that General Washington's headquarters were at Mrs. Wallace's about twelve miles from New Brunswick." Married senior officers were joined by their spouses and children once they settled into their quarters.

The generals occupied every available house in the area and were required to pay rent. Five historic houses near the Middlebrook Campground served as headquarters for the top officers of the

Continental Army. All have survived and have been painstakingly restored. Washington's headquarters was at the Wallace House, Baron von Steuben lived at the Abraham Staats House in South Bound Brook, and General Henry Knox lived at the Jacobus Vanderveer House at Pluckemin, near the artillery cantonment. Lord Stirling set up headquarters in Bridgewater at the Van Horne House on Main Street, while Quartermaster General Nathanael Greene had his headquarters at the Van Veghten home, located near the bridge over the Raritan River south of the camp. The houses offer a unique historical perspective on the military encampment. All now serve as offices for local historical societies and are open for public tours.

The senior officers often competed among themselves for the proper available living quarters. Evidence of disputes in seeking housing is first found in General Greene's correspondence. Colonel Beatty, Commissary General for Prisoners, asked Greene to evict Chaplain Alexander Balmain from a house. Beatty requested that Greene order Balmain to be "removed to a greater distance where he might be equally well accommodated. The duty of my office requiring a constant attendance to the Commander-in-Chief, I cannot find with the circuit of several miles any house convenient"[1]

Joseph Dunn, Chief of Express Riders, returned to his house and found that it had been confiscated by General William Woodford. Dunn complained to Greene that he found that this was "very disagreeable." He returned later and threatened Woodford for his eviction.

Apparently, most homeowners graciously accepted the confiscation of their homes and lived amicably with officers and their staffs. Only one instance of a difficulty between homeowners and officers can be found. General Anthony Wayne and his staff occupied the Abraham Van Nest House in Millstone. A verbal confrontation between Wayne's staff officers and the local constable and Abraham's son caused such a commotion that local officials reported the incident to General Washington and Governor Livingston.[2]

Wallace House

When selecting a home for his headquarters at the winter camps General Washington typically avoided ostentatious dwellings, and instead selected large farmhouses that could accommodate his staff and numerous visitors. One or two rooms and an adjacent office were reserved for Martha and himself. During the Middlebrook Encampment he rented the Wallace House, one of the largest homes in the area. He paid the owner $1,000 for four months use of the house and its furnishings.

This Georgian mansion was owned by John Wallace, a retired Philadelphia merchant, and was the largest home built in New Jersey during the Revolutionary War. It was completed in 1776, just two years before the second Middlebrook Encampment. The structure has eight rooms and an arched center hallway that was reserved for the use of the Commander-in-Chief. Washington and his staff shared the home with the Wallace Family. Candlesticks, china and other small furnishings were provided for the general's use by the army. He graciously presented these objects to Wallace when the army departed.

The two-story house had four bedrooms on the upper floor. George Washington and Martha Washington occupied one room. Aides-de-camp and secretaries took another room, while their hosts, John Wallace and his wife, Mary, filled a third room. Mary's mother, Mary Maddox, stayed in the fourth bedroom. A room on the first floor served as a cramped office for Washington and his staff.

Today the restored house stands at 38 Washington Place in Somerville, six miles from the center of the regimental campsites, and a safe distance behind what could potentially become front lines of defense. The home is situated just south of Old York Road which paralleled the north bank of the Raritan River. The route was the main road from New York to Philadelphia. John Wallace wrote a receipt at Raritan on June 4, 1779, "Received of Major Gibbs one thousand dollars for the use of my house Furniture &c. &c. Which His Excellency General Washington had for his Headquarters."[3]

The Wallace House, Somerville, NJ. *(wikipedia.org)*

General Washington moved into the house on December 11, 1778. After spending only eleven days at his new headquarters, he left for Philadelphia to attend Congress for six weeks. He returned in February 1779 with his wife Martha, and accompanying servants and aides. Martha remained with him at the Wallace House until June 3, 1779, when the couple left Middlebrook with the army.

Washington and his staff were busy during this time hosting foreign dignitaries, preparing dinner parties, and making plans for a spring military campaign. Six aides-de-camp assisted with the correspondence required to manage all the operations of the military. Overcrowding in the Wallace house must have made living uncomfortable. Aside from Martha, who spent the entire winter with her husband, Washington's aides-de-camp Alexander Hamilton, James McHenry and Tench Tilghmam also lived at the house and shared a small bedroom. Dinner was always a considerable affair, with as many as thirty high-ranking officers or visiting dignitaries attending.

While at the Wallace House, Washington and his staff planned and gathered supplies for the successful 1779 large scale campaign in New York State. This operation against the Native American allies of the British was launched in June and involved almost a third of the

Continental Army including the New Jersey Brigade. It was led by Generals John Sullivan and James Clinton.[4]

A household staff lived nearby. They cleaned, cooked, and performed housekeeping chores for the occupants, as well as doing the same for Washington's Life Guard. This unit of about 200 troops was stationed close by in Raritan. The Commander-in-Chief's Guard was authorized on March 11, 1776, soon after Washington arrived at Cambridge, Massachusetts to take command of the American forces there. The purpose of the Life Guard was to personally protect the general, in addition to the funds and official papers of the Continental Army. Washington personally created this headquarters security detachment and directed that it consist of a "corps of sober, intelligent, and reliable men." Because it was an honor to belong to the select unit, care was taken to ensure that soldiers from each of the thirteen states were represented.

The Wallace House became a museum in 1897, thanks to The Revolutionary Memorial Society, a group composed of New Jersey citizens. Public outcry about the need to preserve this historic site, combined with assistance from the Frelinghuysen family, led to the salvaging and preservation of the home and the adjacent Old Dutch Parsonage in 1913. Both structures became New Jersey State historic sites in 1947 and were added to the National Register of Historic Places on December 2, 1970. Today the Wallace House and Old Dutch Parsonage Association continue this legacy of citizen involvement by maintaining the homes as furnished house museums.

The Abraham Staats House

This house is one of the finest surviving examples of the Dutch settlement of the Raritan Valley in the 18th century. It stands on Von Steuben Lane off Main Street in South Bound Brook. It was three decades old at the time of the American Revolution. The original structure, two rooms and a garret, was built about 1740 on 305 acres

The Abraham Staats House, South Bound Brook, NJ. *(wikipedia.org)*

owned by the Staats, a Dutch family. The family occupied the home for two centuries, and expanded it to its current size.

Abraham Staats, an ardent Patriot, lived here with his family during the Revolutionary War. The British were aware of his political affiliation and specifically excluded him from the general pardon that was offered to rebellious Americans who would declare loyalty to the king. Staats was a farmer and a surveyor, who also served as a tax collector and county freeholder. The main British column passed the home and pillaged it enroute to the Battle of Bound Brook on April 13, 1777.[5] Staats filed a detailed report in 1782 to claim reimbursement for wheat, clothing, and farm animals that were stolen during the raid.[6]

During the Middlebrook Encampment, the Staats family hosted General Baron von Steuben, a Prussian officer who claimed to have served under Frederick the Great during the Seven Years War. He met Benjamin Franklin in Paris while the American diplomat was arranging for French support for the Patriot cause, and Franklin recommended that von Steuben be appointed to train the raw American Army in battlefield tactics. During the Valley Forge encampment, the baron wrote a comprehensive training manual.

Von Steuben's experience with European military tactics enabled him to convert untested American recruits into an effective fighting force. He was appointed Inspector General of the Continental Army on May 5, 1778, and he arrived at the Staats House on March 26, 1779, during the second Middlebrook encampment. By late spring, his systematic drilling and preparation at the camp completed the molding of the Continental Army into a professional fighting force. Washington showcased his now well-organized army at a grand review on May 2, 1779.

The event was held to honor the French minister and a Spanish diplomat. Four battalions performed precise military formations to demonstrate their mastery of von Steuben's training. After the review, about sixty generals and colonels attended a dinner hosted by von Steuben in a large tent set up in the yard of the Staats house. During the event, the French minister informed General Washington that the French army and fleet would assist him in the war and that supplies coming from France would be increased. When the Middlebrook Encampment finally ended in June, von Steuben's efforts had created an American Army that was better trained and disciplined than ever before.

The Abraham Staats House was acquired by the Borough of South Bound Brook in 1999 and lies on four acres along the Delaware & Raritan Canal. Grant assistance provided by Somerset County to the Borough has ensured that this important site remains in the public domain. The Friends of Abraham Staats House, a non-profit organization, in cooperation with the South Bound Brook Historic Preservation Advisory Commission, helps support the maintenance of the site. The Battle of Bound Brook is recreated here every year on its anniversary, April 13, after which reenactors camp on the grounds. The Abraham Staats House was entered into the National Register of Historic Places on December 4, 2002.

The Van Horne House

This Georgian Mansion was built on Old York Road in Bound Brook in 1750. It was the country estate of Philip Van Horne, a prosperous

Van Horne House, Bridgewater, NJ. *(wikipedia.org)*

New York merchant. He named it "Convivial Hall" because of the generous hospitality he offered there. Anyone who visited the Van Horne House was welcomed regardless of their political loyalty. During the war the house was constantly full of visitors.

The owner's shrewd motive may have been to find men worthy of his four marriageable daughters. Andrew D. Mellick, Jr. in his book, *The Story of an Old Farm*, reports that Van Horne had "five handsome and well-bred daughters who were the much admired toasts of both armies."[7] More likely, like many other Americans, Van Horne did not want to be on the losing side of the war. George Washington suspected that he was a Loyalist, but still allowed him to occupy the house. General Benjamin Lincoln was living in this home at the time of the Battle of Bound Brook on April 13, 1777. He was surprised by British soldiers charging up to the house and was forced to quickly escape out the back door. That day, Van Horne hosted British General Cornwallis for breakfast and American Generals Lincoln and Greene for supper. After the British victory, the house became the British headquarters at the Bound Brook outpost.

The Van Veghten House, Bridgewater, NJ. *(wikipedia.org)*

During the Second Middlebrook Encampment, General William Alexander, also known as Lord Stirling, used the home as his headquarters. When General Washington left Middlebrook to meet with Congress in Philadelphia, Stirling was left in command at Middlebrook.

On October 26, 1779, British Colonel John Simcoe led a group of the Queen's Rangers to destroy American landing boats moored on the Raritan River and to capture New Jersey Governor William Livingston, who was believed to be living in the Van Horne house. He raided several houses in Bound Brook and took several prisoners but did not find Livingston.

John Herbert, who owned a nearby mill powered by the Middle Brook, substantially altered Van Horne's house in the 1830s. It was later rescued from neglect by the Calco Chemical Company in the 1930's and restored in the Colonial Revival style for use as its offices. The house is currently owned by Somerset County. Since 2002, the

Heritage Trail Association has used it as their headquarters. It was added to the National Register of Historic Places on March 8, 2002.

The Van Veghten House

The Derrick Van Veghten House stands on its original site on the north bank of the Raritan River. It is on high ground overlooking the meadows on the riverbank. For the most part, the setting still appears as it would have looked during the Revolutionary War. Van Veghten owned a 1000-acre tract extending from the river to the mountain. The house, in Bridgewater, is about three quarters of a mile south of Old York Road, near Finderne Avenue.

The present structure evolved from the first house built by Derrick's father Michael Van Veghten before 1720. The use of brick indicates that he was an affluent man, since most houses in the Raritan Valley were wood frame. The masonry pattern formed by laying bricks long ways and crossways, known as Flemish bond, is an indication of the owner's Dutch heritage.

The house served as the headquarters for Quartermaster Nathanael Greene during the Middlebrook Encampment in 1778-1779. Van Veghten offered his entire property for the use of the Continental Army. Parts of the Pennsylvania regiments were encamped in his fields at the time. The staunch Patriot never asked to be paid for the trees which were cut down to build huts, or for those burned for firewood.

The Greenes entertained most of the prominent people who visited the camp during that winter. Greens vivacious wife Kitty was the most popular woman there during the social season. While in residence, General Greene wrote a letter to Jeremiah Wadsworth describing "a pretty little frisk" held at the house on March 19, 1779. Throughout the course of the evening, General Washington danced with Mrs. Greene "upwards of three hours without seting down."[8]

During his stay at the house Nathanael Greene was assigned

the difficult job of supplying the army. He faced shortages of food and forage, and difficulties procuring horses and wagons for transportation over impassable roads to the Middlebrook Encampment. Inflation of the Continental dollar severely restricted his purchasing power.

After Derrick Van Veghten's death in 1781, the house passed through several families. It was renovated several times between the 1850's and the early 20th century, and was purchased in 1934 by the Singer Company. In 1971, the house and its one acre of land were deeded to the Somerset County Historical Society by the Singer Company. It was placed on the National Historic Register in 1979. The Van Veghten House has been restored and now serves as the headquarters of the Somerset County Historic Society.

Headquarters for Other Field Grade Officers

General Knox was headquartered in the Jacobus Vanderveer House, which is described in the Pluckemin section of this book. Other officers lived in private homes in the Bound Brook and Middlebrook area while their huts were being built. Their quarters were usually shared with the owner's family. Field grade officers, the highest-ranking commanders, were permitted to live outside of the camp. This group included the commander-in-chief, six major generals, six brigadier generals, and eighty-five colonels and lieutenant colonels. After the huts were completed all others were ordered to report back to the campground to live with their regiments. Their presence among the troops was believed to be necessary to maintain discipline. Other than the restored historic houses described above, only three others that billeted senior officers have been identified. These homes have all been demolished in past years.

Major General Baron Johanne DeKalb, commander of the Maryland Division, was headquartered near the Queen's Bridge that spanned the Raritan River at Bound Brook. This house was about a quarter of a mile up the road from the Staats House. It stood near the intersection of Main Street and Washington Street in South Bound Brook. The location allowed DeKalb to remain close to his regiment.[9]

Major General Louis DuPortail, Chief of Engineers, usually referred to as sappers and miners, stayed at the house of a Mr. Baker in Middlebrook. The house has been torn down and its location has been lost to historians. Lacking trained engineering officers, the American's relied heavily on foreign officers, mostly from France, for sorely need technical assistance. The Engineers Divison was new and was still in the process of being staffed. Its troops came from the state line regiments.[10]

Major General Arthur St. Clair, Pennsylvania Division, replaced General Anthony Wayne in February 1779 as commander of the Pennsylvania Line. He was housed near his troops on Millstone River Road where Central Jersey Regional Airport is located today. General John Sullivan also stayed there with St. Clair. Together they planned the campaign against the Iroquois Nations which began later that year. The headquarters for adjutant generals was located on the south side of the road from the Wallace House, near Van Middleswoth's Island. The house and the island no longer exist.

The magnificent restored historic homes that housed senior commanders of the Continental Army during the Second Middebrook Encampment provide a special insight into both their military and personal activities during that time. Each one of these superb surviving structures has a story to tell. Researching the lives of private owners of the homes also uncovers many details of the social history, political affiliations and culture of that critical time.

Chapter Twenty-One

Washington on the Rocks

Green Brook, Middlebrook, and Other Perches of the American Eagle

GENERAL WASHINGTIN VISITED several viewpoints along the first Watchung Ridge to observe the British troop movements and to plan his own maneuvers. Tracing the history of the famed Washington Rock in Green Brook Township, New Jersey the most well- known of these lookouts, led to a fascinating discovery. A reference to another rocky observation post along the ridge appears in a British account penned in 1785. This document mentions Washington's observation position as being on a rock on the south side of Middlebrook Heights. This would place it in the Middlebrook Encampment on the first ridge of the Watchung Mountains, about two miles southwest of Green Brook.[1]

General Washington shifted the American Army from Morristown to Middlebrook after learning that British forces under Howe might be preparing to move across New Jersey to capture the American capitol at Philadelphia. American troops began arriving in May 1777, and they remained until early July. This visit, taking place over a tumultuous seven weeks in the spring of 1777, was a time of uncertainty and indecision for the Americans. It has become known as the First Middlebrook Encampment.

From Middlebrook Washington could move his army quickly down from the security of the Watchung Mountains to strike the flank of the enemy columns if they attempted to move anywhere in New Jersey. What would be the next move of the mighty 16,000- man British force billeted around New Brunswick? Would they head south to take Philadelphia? Would they relocate west to occupy all of New Jersey? Or would they withdraw and cross over the Arthur Kill to Staten Island in order to return to New York?

It would be logical for General Washington to establish an observation post on the Middlebrook Heights. From here British activities in the New Brunswick area could be observed, and the lookout also provided a sweeping view from that town, seven miles away, all the way to the Amboys. Enemy activity could be observed on the Raritan River and on the road to Pluckemin to the west. The other lookout, at the iconic Green Brook Rock, two miles north, was more distant and would have value only if the Redcoat Army moved north in that direction.

Various references over the years claimed that the Middlebrook Rock was actually Chimney Rock, a natural stone outcrop that resembles a chimney stack. This landmark has been well-known throughout the state's history, but the view from there is down a valley and is restricted by hills on each side. Because the view is limited it can be ruled out as being the Middlebrook Washington Rock. Chimney Rock, identified on current maps as Hawk's Watch, is a mecca for bird watchers today.

Another early reference to the Middlebrook Rock was from local historian Rev. Dr. Abraham Messler, pastor of the nearby First Church of Raritan. He described the rock outcrop in the 1830s but did not disclose its exact location. He reported, "On the apex of the Round Top, on the left of the gorge in which Chimney Rock stands, there are yet to be seen the rude remains of a hut which Washington sometimes frequented during those anxious months of 1777. On the east side of the gorge, also, fronting the plain north of Middlebrook, there is rock which has been named 'Washington Rock,' because there he often stood to gaze anxiously upon the scene it overlooks."[2]

The exact location of the Middlebrook Washington Rock was lost to history. Only a few vague references to a site somewhere along the ridge are in documents written between Messler's visits in the 1830s and 1975, when A.A. Boom reported that the location of the rock was mentioned in a book by historian Benson Lossing, who had actually visited the site in 1851.

Lossing's *Pictorial Field-Book of the Revolution* was published in 1853. To gather material for his three volume narrative sketchbook of the American Revolution, he traveled some 8,000 miles to visit sites in the United States and Canada. When Lossing arrived in New Jersey he asked to be taken to "Washington Rock." At the time the local residents assumed that the rock was the rocky ledge above the Middlebrook Encampment near Chimney Rock, not Green Brook. While Lossing mentions the Green Brook Rock local residents apparently regarded it as less historically important than the Middlebrook lookout.

A search of Lossing's journal provided a mother lode of information. Lossing stated that the rock was at the end of the old steep road over the mountain. Not only did he describe the location in lucid detail, but he also drew a sketch of the rocky ledge which was identified by locals as "Washington Rock." Lossing's account of his visit to the Middlebrook Rock provided enough detail to find the approximate location using existing landmarks:

> *Returning to the village, we proceeded to visit the campground, which is upon the left of the main road over the mountains to Pluckemin; also "Washington's Rock." The former [at Green Brook] exhibits nothing worthy of particular attention; but the latter, situated upon the highest point of the mountain in the rear of Middlebrook, is a locality, independent of the associations which hallow it, that must ever impress the visitor with pleasant recollections of the view obtained from that lofty observatory.*
>
> *We left our wagon at a point half way up the mountain, and made our way up the steep declivities along the remains of the old road [Vossellor Avenue]. How loaded wagons were managed in ascending or descending this mountain road is quite inconceivable, for it is a difficult journey for a foot-passenger to make. In many*

places not even the advantage of a zigzag course along the hillsides was employed, but a line as straight as possible was made up the mountain. Along this difficult way the artillery troops that were stationed at Pluckemin crossed the mountain, and over that steep and rugged road heavy cannons were dragged.

Having reached the summit, we made our way through a narrow and tangled path [Miller Lane] to the bold rock.... It is at an elevation of nearly four hundred feet above the plain below, and commands a magnificent view of the surrounding country included in the segment of a circle of sixty miles, having its rundle southward.

At our feet spread out the beautiful rolling plains like a map, through which course the winding Raritan and the Delaware and Hudson Canal. Little villages and neat farmhouses dotted the picture in every direction. Southward, the spires of New Brunswick shot up above the intervening forests, and on the left...was spread the expanse of Raritan and Amboy Bays, with many white sails upon their bosoms. Beyond were seen the swelling hills of Staten the Island, and the more abrupt heights of Neversink or Navesink Mountains, at Sandy Hook. Upon this lofty rock Washington often stood, with his telescope, and reconnoitered the vicinity. He overlooked his camp at his feet, and could have descried the marchings of the enemy at a great distance upon the plain, or the evolutions of a fleet in the waters beyond.[4]

Lossing's route can be traced on a 1777 map drawn for General Washington by Captain William Scull.[5] This map shows the locations of the Continental Army units during the First Encampment in 1777, as well as both Vosseller Avenue and Chimney Rock Roads. An icon marks the exact location of Wayne's Regiment, which occupied the hilltop. The location of the rock could now be closely approximated.

With Lossing's sketch in hand, I left Route 22 West to drive up the winding hill on Vossellor Avenue to the crest of the first ridge. After turning left onto Miller Lane, I began looking out for the rocky ledge

Washington Rock, Middlebrook, as drawn by Benson Lossing 1851.
(The Pictorial Field-Book Of The Revolution)

depicted in Lossing's sketch. A sign on the left side of Miller Road read: Eagle's Nest Museum-Herbert M. Patullo.

Nearby was a large barn and a private home. The barn appeared to contain antiques. Adjacent to these structures was a grassy plot with a flagpole that overlooked a breathtaking panoramic view. I knew at once that this had to be vista described by Lossing.

I was soon greeted by Herb Patullo, owner of the premises. He examined Lossing's sketch of the rock outcrop and noted that it looked very familiar and might be close by. We tramped through the woods for only about fifty yards, and soon found a rock outcropping which closely resembled Washington Rock at Middlebrook, as drawn by Benson Lossing in 1851.

Herbert M. Patullo was the owner of the small Eagle's Nest Museum that is housed in the barn. He told me he has lived his entire life in Bound Brook and for many years was a restaurateur in the town. He has served the community as a leader in many cultural and historical community activities, among them as president of the Washington Campground Association. He was an astute local historian with a prolific knowledge of the history of the Bound Brook area and the Middlebrook Encampment.

As a child his father instilled in him an awareness that their town and the heights above it were the setting for many critical events during the American Revolution. This understanding led to his lifelong interest in the encampment, and in efforts to preserve the historic site from desecration by commercial and residential development. Herb Patullo passed away in 2020. His obituary recounts his interesting life, including his dedication to the Middlebrook Campground:

Herb Patullo at Middlebrook Washington Rock in 2019. *(Author's Collection)*

> *Bound Brook - Herbert (Herb) Patullo entered into Eternal rest on Saturday, August 15, 2020. Born in Bound Brook, Herb lived there all his life. He was the son of Marianna and Benjamin Patullo. Herb was a successful business owner, who along with his family, ran the very popular restaurant known as Patullo's.*
>
> *Herb left high school at the age of 16 with the intention of serving his country by joining the military. However, being too*

young to join the service at the time, he was offered employment at the Army Supply Depot in Somerville. He worked there as a food handler for 2 years during World War II. After the end of the war, Herb enlisted in the Navy serving on the destroyer, Charles R. Ware as a machinist mate. He served 45 months on sea duty before sustaining a back injury. Upon his return home from the Navy, Herb started a landscaping business. He eventually renovated the grocery store that was owned by his family, into a neighborhood tavern which grew through the years, evolving into a full restaurant, nightclub and banquet facility. He was always a very imaginative and successful businessperson.

Herb was one of the originators of the John Basilone parade held each September in Raritan, serving as the liaison between the committee and the Marine Corps band. He also worked on the committee that was responsible for getting a United States postage stamp in honor of John Basilone.

Herb was a supporter of various organizations including veterans' groups, scouting, and the National Museum of the American Revolution, and held lifetime memberships in American Legion Post 63 Bound Brook, Martinsville VFW Post 1388 and a member of Tin Can Sailors.

He was a supporter of the Marine Corps Law Enforcement Agency. He was awarded the Martinsville Am Vet award in 1987, Citizen of the Year by the Bound Brook Elks in 2005 and in 2008 was the recipient of the prestigious Chapel of Four Chaplains Legion of Honor award. Herb was a huge supporter of the community and especially of family. He will be sorely missed.

Herbert Patullo's property on the heights overlooking Bound Brook is today in the town of Martinsville. It is in the heart of the First Middlebrook Encampment, a natural fortress where artillery could guard the pass that is now Chimney Rock Road. The site provided a view of the slice of the countryside that General Washington needed to observe the activities of the British in New York and New

Brunswick. It also was a position from which he could take the offence and attack spread out Redcoat columns if an attempt was made to cross New Jersey to reach Philadelphia. In many ways the place is more significant than other renowned Continental Army camp sites such as those at Valley Forge and Jockey Hollow.

The Pennsylvania Brigade camped in an area that extended a half mile from the intersection of Vosseller Avenue and Hillcrest Road, along Miller Lane towards Chimney Rock. Washington placed his best armed and trained brigade in this position of honor on the front line, in front of his main army. Commanded by General Anthony Wayne, it was recognized as an elite fighting unit. This position secured the strategic road passes through the mountain and guarded the rest of the army that was spread out in the Washington Valley behind the ridge. Wayne's Brigade was made up of four Pennsylvania Regiments. They were armed with .69 caliber Charleville muskets from France. Many of these Pennsylvanians were veterans of the Trenton-Princeton campaign and earlier operations in New York State.[6]

Visitors are attracted by the panoramic view from Wayne's camp along the first ridge, but most do not appreciate the historical significance of the site. The land had been owned by the adjacent (and still active) Stavola basalt quarry since the late 19th century. In the early 1970s the property was offered for sale, its only structure being a single dilapidated building that housed a county home for the indigent. Herb Patullo became concerned that encroaching commercial and residential development would soon envelop this historic place. When the county home was closed in 1974, he purchased the building and the three-acre lot on which it stood. Later, in 1988, he acquired the entire forty acres along the ridge between Chimney Rock Road and Vossellor Avenue. This large swath enveloped the entire hilltop of the Revolutionary War campsite. He built his home and barn on the site twelve years later, and eventually sold most of the property back to Somerset County. The county added this historic land to the adjoining Washington Valley Park in 1994.

The Eagle's Nest Museum on Patullo's land is now closed. It

contained a modest eclectic collection of historic memorabilia, as well as items from Patullo's personal life. Two original paintings, by local artist Victor Temporra, are on display in his former home. They portray scenes of the camp during the time it was occupied by Washington's Army. Many of the details depicted in the paintings were provided by Patullo himself.

McBride's Rock

As the speaker at the annual meeting of the Washington Campground Association in February 2017, I took the opportunity to describe the possibility that the rock at Middlebrook on the Patullo property was the observation post along the ridge mentioned in the British account of 1785, and also the one visited by Lossing in 1851. Afterward, I was contacted by local historian Don McBride, who claimed to have found another observation rock that more closely matched Lossing's sketch. Its location was only a thousand yards north of the Patullo Rock, and nearer to Vosseller Avenue.

McBride lived on the ridge on Hillcrest Road, adjacent to the former campground. He has explored the area on foot for many years and identified remnants of trenches and stonewalls and has also cleared paths to the crest of the ridge. I joined him to visit the Vosseller Avenue rock a few weeks after our first contact. We tramped through the foliage to the edge of the cliff and stood on a rocky outcrop, and enjoyed the same sweeping panoramic sixty-mile vista. We then descended about fifty feet down the hill to the base of the cliff to view the lookout from below. We then compared this rock to the one on Patullo's land and tried to match both with Lossing's sketch and narrative.

McBride's Vosseller Avenue rock formation appears to look much more like Lossing's drawing. Its location atop a distinct cliff resembles his sketch and the site is closer to Vosseller Avenue, as described by Lossing. In addition, the site is also exactly where the Wayne

Brigade was camped. Its elevation is 375 feet, 65 feet higher than the Patullo Rock.

However, the view from Petullo's Eagles Nest is superior to the southeast toward New Brunswick, if trees could be removed. Moving down the ridge a hundred yards toward Petullo's home allows an improved view from New Brunswick to the Sandy Hook area. It is likely that General Washington used both places along the ridge between Vosseller Avenue and Petiullo's Eagles Nest in 1777 to detect the movement of his Redcoat adversaries. Both sites qualify are equally qualified as Middlebrook Washington Rock.

The Legendary Washington Rock at Green Brook

Washington Rock at Green Brook on the first Watchung ridge overlooks that town and the adjacent city of Plainfield. It is two miles north along the ridge from the Middlebrook rocks. This iconic lookout, now a state park, is commonly recognized as the only "Washington Rock" in New Jersey.

After the war, in the early 1800s, this rock outcrop became a popular tourist attraction. About 2,500 area residents visited the site on July 4, 1831, to celebrate the 75th anniversary of the nation's independence. To accommodate the influx of visitors, a road called Cardinal Lane was constructed for stagecoaches to shuttle tourists between the Plainfield Railway Station and the rock. Families picnicked on the grounds near the rock, and in the years that followed inns and hotels were opened there. Cardinal Lane remains today as an unpaved hiking trail. Community leaders attempted to recognize Washington Rock with a monument. This proved difficult as ownership of the property frequently changed.

The earliest public reference to the Washington Rock at Green Brook was in 1844, sixty years after the Revolutionary War. Barber and Howe's Historical Collections of the State of New Jersey provides this description:

General Washington with artist Charles Willson Peale. View of the Battle of the Short Hills June, 26, 1777. *(Sketch by Peale from* The Selected Papers of Charles Willson Peale and His Family)

At an elevation of about 400 feet on the brow of the mountain in the rear of Plainfield stands Washington's Rock. It is one of very large size, being about 25 feet in height and from 30 to 40 in circumference. The bold projection, which nature has given it from the surface of the eminence, renders it a fine position for taking an extensive view of the country below.

In June, l777 the American Army was stationed at various places on the plains below. After the retreat of Sir William Howe from New Brunswick, Washington retreated to the heights rather than confront the enemy. The advance guard of the British army led by Cornwallis fell in with Lord Stirling's Division on June 27, 1777.

At various times he [General Washington] resorted to this place to ascertain the movements of the enemy. This circumstance has given the Rock a sacred character to the people of the present day, which, in connection with the beautiful prospect it affords, has made it a place of resort for parties of pleasure.

The scene is one of uncommon beauty. The whole country, apparently, lies as level as a map at the feet of the spectator, for a circuit of 60 miles. On the left appear the spires of New York City, part of the bay, Newark, Elizabeth, Rahway and New Brighton [Edison]. Directly in front are Amboy and Raritan Bays. To the right New Brunswick and the heights of Princeton and Trenton; and far to the southeast the eye stretches over the plains of Monmouth to the heights of Navesink. Beautiful villages bedeck the plain; and cultivated fields, farm houses and numerous groves of verdant trees are spread around in pleasant profusion.[7]

The claim that General Washington actually used the rock as an observation post was verified in 1897 when evidence of an eyewitness account was found. Sometime before his death in 1830 this account was related to George W. Fitz-Randolph by two local farmers, Ephraim and Josiah Vail. Fitz Randolph, a descendant of the Vail family, owned several farms below the Rock in Green Brook.

In the year 1777 or '78 Washington, with 6,000 men, was encamped on the Ridge at Middlebrook near and west of Bound Brook. The British army were encamped at New Brunswick, Rahway and Perth Amboy, making incursions into the surrounding country. Doubtless, with an intent of guarding against a serious incursion or surprise, Washington was on his way to the top of the mountain back of Green Brook.

Be that as it may, he, with an aide-de-camp, mounted, rode in the gateway and up to a group of men standing between the house and the barn on the farm, now known as the Jonah Vail farm. Washington said: 'Can any of you gentlemen guide me to some spot on the mountains from whence a good view of the plain below can be obtained?' Edward Fitz-Randolph, one of the group, said: 'I know of the best point on the mountains for that purpose' and added that, if he had his horse, he would take him to it. Thereupon the General requested his aide to dismount and await his return. Fitz-Randolph, mounted upon the aide's horse, piloted the General to the Rock, which today bears the historic name of 'Washington's Rock.'

I have given the above nearly word for word, as given to me by Ephraim Vail, who died a few years since aged 90 and over, on the farm where he was born and raised. Josiah Vail gave me the same version of the incident; indeed any of the old residents of Green Brook would corroborate the same, were they alive. All these Vails were Quakers, owning adjoining farms, and their word is reliable.[8]

George Fitz-Randolph again confirmed Washington's presence at the Rock shortly before he died. He reported that Washington observed the British Fleet preparing to sail to Chesapeake Bay: "Looking through his glass, Washington rejoiced at finally watching the British fleet of 270 transports leave the bay at Amboy and head to sea and leaving Jersey forever."

The development of the rail line below the rock in 1838 accelerated the development of the towns below. Westley H. Ott, the unofficial historian of Dunellen, New Jersey, reported that the rock was a popular spot for political rallies as early as 1840. Rail service was provided between Elizabethtown and Plainfield at that time and was later extended to Bound Brook in 1843.[9]

The rock has long been a landmark and site for day trips for Central Jersey residents. As early as 1840, residents from nearby towns began climbing up to the rock to admire the view and have picnics. Rebecca Vail, a Quaker farm wife of Green Brook, kept a diary for a brief period in her life. That fragment of journal records that in 1847 she made a picnic trip to the rock.[10]

Historian Ott also reported that on the 75th anniversary of Independence, in 1851, more than 2,500 spectators visited Washington Rock. He also described the Washington Rock Mountain House built in 1852, as a "large three-story building" with a full-length porch. The hotel later burned to the ground in a raging fire in 1883.

In 1851, historian Lossing provided this brief mention of the

Green Book Rock after his lengthy detailed description of the large stone lookout at Middlebrook which he apparently considered to have more historical importance:

> *In the rear of Plainfield, at an equal elevation, and upon the same range of hills, is another rock bearing a similar appellation, and from the same cause. It is near the brow of the mountain, but, unlike the one under consideration, it stands quite alone, and rises from a slope of the hill, about twenty-five feet from base to summit. From this latter lofty position, it is said, Washington watched the movements of the enemy in the summer of 1777."* [11]

An 1860 real estate map titled "Washington Rock to Newark Bay," part of the collection of the Cannon Ball House in Scotch Plains, shows two hotels at the site. In the 1870s, a stagecoach regularly ran two miles between the Dunellen Station and the rock. After the Civil War, John W. Laing, a Plainfield stableman, began climbing the mountain each year to give the rock a coat of whitewash.[12]

The Constitutionalist, Plainfield's leading newspaper from 1868 to 1911, reported that painting the rock had become a notable annual event.[13] In 1898 The Constitutionalist proclaimed that Washington Rock was to be defended, not against the assaults of a Spanish Army, but by the ravages of time.

> *It is really but right that that historical spot should be looked after especially at such a time as this. For years, it has been the custom of some patriotic citizens to whitewash the Rock so that it can be seen for miles around. Constable William N. Pangborn and Edward Conshee are leading a group to ensure that the Rock receives a coat of whitewash and the brush cleared around it.*[14]

In 1867 a group of citizens formed the Washington Monument Historic State Association to prepare for the centennial in 1876.

The group included many members of the local Masonic lodge in Plainfield. George Washington was an active Mason all his life and served as the Grand Master of the fraternity. The association began raising funds to build a 100-foot observation tower on the mountain behind the rock. The tower, was to be "a monumental shaft dedicated to immortal Washington which greets the rising sun from yonder mountain brow." The Masons provided the inscription—"New Jersey gratefully remembers her defenders in the dark and bloody days of the Revolution."[15] Although the foundation and cornerstone were laid, mortgages held on the land prevented the association from getting a clear title. The project was eventually abandoned.

While the observation tower was never built, ceremonies, parades, speeches, fireworks and displays held on July 4, 1876, were extensive. A band provided music all day at the huge dancing platform constructed at the rock. In the evening the grove and the hotel were brilliantly illuminated, and music continued for dancing, followed by a grand Centennial Supper for 6,000 people.

Interest continued in erecting some type of memorial at the rock. Eventually the Green Brook Daughters of the American Revolution (DAR), Continental Chapter, raised funds in 1879. A.L.C. Marsh, who lived in Plainfield, was a New York City architect who designed "country homes." He donated a memorial design plan to the DAR which proposed building a stone cairn on the foundation of the tower built forty years prior by the Masons. A native stone retaining wall to join the two rocks on the site would also be constructed. This structure was never built due to funding difficulties. In 1909 the property was purchased by Mr. Charles McCutchen of North Plainfield, who understood its historical significance. He held it in trust for the people of Plainfield and North Plainfield.

Finally, in 1912, a small cube-shaped, one room stone building was erected by the DAR at the cost of $3,000. This monument, surmounted by a high flagpole, stands today approximately twenty feet above the rock. On March 27, 1913, the New Jersey Senate and State Assembly passed an act to acquire Washington Rock and adjoining lands as a gift from McCutchen. The act also called for the appointment of a commission to improve and maintain the site as a public park.

Today, Washington Rock Park, with the DAR monument, encompasses two large rock outcrops about eighty feet apart. One is called "Lafayette Rock" to honor the young French general who accompanied Washington, and who is reported to have perched on it in 1777. The park is currently operated and maintained by the New Jersey Division of Parks and Forestry.[16]

Other New Jersey Washington Rocks Along the First Watchung Ridge

Another Revolutionary War lookout post was located above Montclair State University at the intersection of the Great Notch area of Little Falls, and the Montclair Heights section of Clifton. Eagle Rock in West Orange served as a lookout site, as well. Two other places on the South Mountain Reservation, in central Maplewood, Millburn, and West Orange, are said to be Revolutionary War lookouts. One is marked with two stone pillars.

In June 1780, sentries at these northern lookout stations reported the dreaded news that the British had finally launched an attack westward. The assault moved toward the Hobart Gap, the pass through Chatham and Madison, now Route 24. This corridor led west to Jockey Hollow, the Morristown winter encampment of the Continental Army. A breakthrough would have been devastating to the debilitated army while it recovered from the most brutal winter ever recorded.

Five thousand enemy troops advanced through Union Township, then called Connecticut Farms. A British column rushed along Galloping Hill Road while Hessian troops under the command of Lieutenant General Wilhelm von Knyphausen pressed up Vaux Hall Road. This joint attack was vigorously repelled at Connecticut Farms by the Continental Army and New Jersey Militia forces under the command of General William Maxwell and at Springfield on June 23, sixteen days later. The warnings provided by as many as twenty-three observation posts along the first Ridge of the Watchung Mountains prevented an American defeat and an early end to the War.

Chapter Twenty-Two

Scarce Research

Ambitious Archeology and a False Start

ARCHEOLOGICAL STUDIES AT sites on the Middlebrook Campground have revealed much about what actually occurred there during the two encampments. They also confirm much of its scant written history. Various diggings have turned up remnants of huts and fortifications that had been ploughed over or were enveloped by dense woods. Excavations have also exposed unmarked graves, bullet or cartridge cases, fragments of clothing, traces of lost roadways, vanished buildings, and lines of earthen fortifications.

Archeology at Middlebrook has verified regimental campsites, troop movements, evidence of struggles with supply shortages and treatment of the sick and wounded. Professional excavations in the main campground that encompasses the greater Bridgewater area, and at the artillery camp seven miles away at Pluckemin, offer insights into the everyday lives of soldiers living at the camp. This confirmation has served to dispel long believed myths regarding activities and composition of the Continental Army in 1778-1779.

The heart of the encampment in the late spring and early summer of 1777, the period of the first Middlebrook encampment, was in the Washington Valley. This area lies along the east and west branches of Middle Brook, on either side of present-day Washington Valley Road. Wayne's camp, along the crest of the first Watchung Ridge, was a small but strategic forward position of this main camp. The second Middlebrook Encampment spread out over about ten miles in the lowlands of the Raritan Valley. It was more extensive and lasted seven months—between late November of 1778 and early June of 1779.

Middlebrook began attracting archeological attention in the early 1970s. Interest was focused on the site of the 1777 Wayne's Brigade encampment. This area in Martinsville extends about a mile, along Miller Lane between Vossellor Avenue and Chimney Rock. When investigations took place, it was undeveloped woodland where archaeological remains might be expected to survive. Most of the land here was publicly held and accessible. However, the surrounding areas on both side of the ridge, along the present-day Route 22 and Washington Valley Road, had been subject to heavy development.

Many intact parcels of land that retain archaeological potential still remain within the ten-mile encampment area. These include tracts of undeveloped woodland, open space, the yards of private homes, and commercial property that has escaped destruction by earth moving machinery. Hints of archaeological potential are occasionally reported when Revolutionary War artifacts are found by random amateur metal detecting.

Early Archeology

The Middlebrook Campground faded into obscurity after the last troops departed in June 1779. No record exists of any interest in archeological study there until the late 1960s. At that time amateur metal detecting began turning up military artifacts on the site of Wayne's 1777 encampment, where Wayne's Regiment of about 600 men camped along the mountain top. A.A. Boom, a local resident, began to explore the ridgeline for evidence of the Revolutionary War camp. He summarized his findings in 1975 in *North of the Rariton Lotts-A History of the Martinsville, New Jersey Area*.[1]

This is where a 1778 Erskine-DeWitt army map shows trenches, cannon emplacements, huts, and fortifications. Boom soon discovered many excavations, old stone walls, and mounds of stone debris in this area. He concluded that the ruins were the remains left after military occupation.

Scarce Research

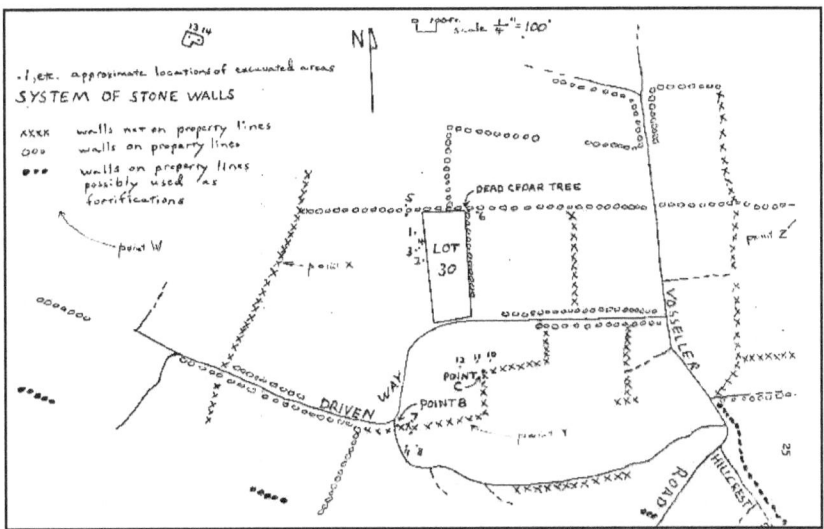

System of Stone Walls and Property Lines, Vossellor Road and Miller Lane by William Liesenbein 1974. Report on a Preliminary Archaeological Investigation of the Alleged Site of the 1777 Summer Encampment of Wayne's Brigade at Middlebrook, New Jersey, June 9 to July 27. *(Courtesy of Department of History, Rutgers University)*

Boom reported that the walls, generally two to three feet high, were originally five feet high, "The size of defense work for the Revolutionary War soldier who had to shoot and load standing up."[2] He found that the walls crossed a large area in roughly the shape of a butterfly along the ridge that straddled Vossellor Avenue, the vital road leading over the mountains to the Washington Valley. Smaller walls ran parallel or at right angles to the main walls. He estimated that the total length of the main wall was approximately three miles and that much of it remained buried.

Boom rejected the idea that the walls marked the property lines of farms. He reasoned that it would have been impossible for farmers alone to move that many tons of stone using their horses to drag it on sleds. He estimated that in 1777 it would have taken a force of 5,000 soldiers about ten days to move the 50,000 tons of stone required for the walls. This indicated that construction by the army was "within the realm of possibility." Based on a 100-year-old map and assorted property deeds, he concluded that the stone walls did not coincide with the property lines or subdivisions of the few farms in the area.

Boom was also convinced that since the entire wall seemed to be identical in size and form this indicated that it had been built "in the same way and time according to a single master plan." He deduced the following: "Weighing the odds, it can be concluded that most of these walls, lacking any other possible purpose, are the long-sought entrenchments mentioned by many historians and referred to in a number of letters to Washington."[3] He did not provide references to these sources.

Boom provided a detailed description and a sketch of the location of the walls in relation to Vosseller Avenue, the Chimney Rock gorge, Miller Lane, and other nearby existing roads. He reported that a number of small redoubts and possible hut foundations could be seen along the smaller walls and "protuberances similar to machine gun nests are located in several places." He observed several depressions about two to three feet deep which he believed could have been fougasse, large land mines created where a large hole is filled with gunpowder and covered with rocks. He also claimed to have discovered camp roads, including the path that Benson Lossing ascended when he visited Middlebrook Washington Rock in 1851.

Boom's Findings Are Challenged

Archeologist William Liesenbein challenged Boom's revealing discoveries in 1974. He excavated five areas where Boom claimed to have found military-related evidence. He concluded that "...the extensive system of stone walls discovered by Mr. A.A. Boom were found to have been built by farmers as field and pasture dividers and/or property lines rather than Continental Army troops as fortifications."[4] Liesenbein found that a main military bastion identified by Boom was in fact a 19th century farmhouse foundation.

Scarce Research 253

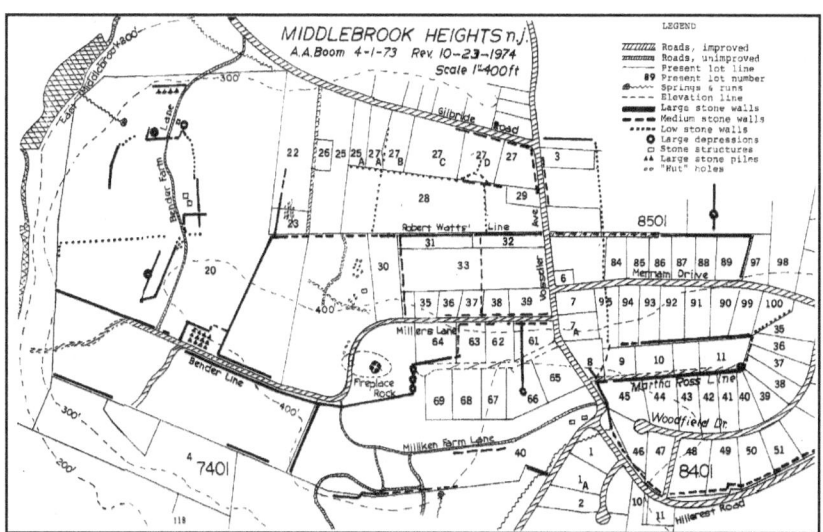

Middlebrook Heights, lots and roads by A. A. Boom. *(North of the Rariton Lotts: A History of Martinsville, NJ Area, Edward J. Maas, editor, Martinsville Historical Committee)*

Stone lime pits that Boom claimed were cooking ovens and garbage pits were natural features, or if manmade had no evidence of being linked to Revolutionary War occupation. Liesenbein agreed that Wayne's encampment was located on the edge of the first Watchung Ridge west of Vosseller Avenue, and that five depressions on the site "could conceivably be the remains of features built by the Continental Army in the summer of 1777." He also concurred that, "concentrations of stones could conceivably have been collected by Continental Army troops in the summer of 1777 in order to clear the ground before pitching their tents." While Liesenbein convincingly refuted many of Boom's discoveries, unfortunately he did little to reveal further evidence of Revolutionary War activity along the ridge.[5]

The visionary A. A. Boom likely erred in his assessment, but he should be credited with focusing attention on the Wayne campsite and the entire Middlebrook Campground. His persistent and inspired efforts eventually led to Somerset County's commissioning of Hunter Research to perform a thorough formal archeological study thirty years later. Boom also left a brief but well written narrative of the

First Middlebrook Encampment in his *North of the Rariton Lotts-A History of Martinsville, New Jersey*.

In 1996 the Cultural Resources Consulting Group conducted an historical and archeological resources study of Washington Valley Park in preparation for sewer alignment for Somerset County. This park consists of 719 acres, and lies along the first Watchung Ridge in Bridgewater Township. The area includes the former Bound Brook Elizabethtown Reservoir and an adjoining park with a network of trails. This report provides an overview of the cultural resources and history of the park from prehistoric times, and describes sites within the Wayne encampment area, but revealed no sites of archeological interest.[6]

A Comprehensive Study by the Elizabethtown Water Company

In 1980 the Elizabethtown Water Company ordered a study for a water pipeline that would enter the Wayne encampment site just south of Vosseller Avenue and continue west along the ridge. A thorough historical and archeological study of the area was conducted by Historic Conservation and Interpretation, Inc. of Newton, New Jersey. Project historian Richard Porter studied several primary sources relating to the encampment, as well as Bridgewater land records during and after the Revolution. After verifying the location of Wayne's camp, he examined stone walls, areas of stone paving, pits, stone piles, and stone-lined depressions that A.A. Boom had described as vaults. Porter agreed with Liesenbien that the stone walls were boundary lines and that other features were of natural origin. None appeared to him to be related to Revolutionary War activity.[7]

Porter's metal detector surveys and excavations of the area extending south from the pipeline to the crest of the ridge revealed interesting consequences from civilian occupation. This land, covered by dense foliage, had once been the site of the fields of the Oliver/McCain/Campbell farm on Vosseller Avenue. A hammer from

Scarce Research 255

a Revolutionary War era, a French flintlock musket and a shard of pottery were discovered. Further digging, below plowed depth, revealed traces of a trash heap containing burnt soil, ashes and bones.

The results of these investigations in manuscript form are kept at the New Jersey Historic Preservation Office in Trenton, along with the hundreds of artifacts recovered from this study. In 2003 there was a plan to combine these materials with those recovered in later years and then turn over the consolidated collection to Somerset County. This plan has still not been implemented.

A Commendable Treasure Hunter

Sadly, over the years, the Wayne encampment area has been the target of intense metal detector exploration by amateur souvenir hunters. This informal removal of countless artifacts has resulted in the depletion of priceless evidence with no record of provenance or attempt at identification. If these objects could have been expertly examined, they could have provided further verification of the boundaries of Wayne's Brigade camp.

There is one report in existence written by a responsible independent treasure hunter. In 1986 Ernest R. Bower, a knowledgeable historian of Revolutionary War history in New Jersey from Bound Brook, found white metal buttons stamped "HR" while searching Wayne's encampment with a metal detector. He describes his excitement of the discovery.

> *Little did I know at that moment that I had made a Revolutionary War relic hunter's find of a lifetime and that thirty feet away from where I was standing I would find an identical button one year later. Carefully brushing off the mud, I could see right away that this was no plain flat button. My heart was beating like a drum as the letters "HR" stared back at me. I had never seen another button like this one. I knew this one was good, real good.*

These artifacts were identified as uniform buttons of Hartley's Additional (HR) Continental Regiment. This American infantry unit served for two years during the Revolutionary War. It was created in January 1777, with Colonel Thomas Hartley appointed as its commander. The corps was comprised of eight companies from Pennsylvania, Maryland, and Delaware. When permanent brigades were formed in May 1777, the regiment was transferred to the 1st Pennsylvania Brigade under General Anthony Wayne. Thanks to Bower, who documented his discoveries, this valuable evidence provided further verification of the Wayne campsite location.[8]

The Hunter Study

In 2003, the Somerset County Park Commission retained Hunter Research, Inc., a Trenton-based consulting firm that specializes in a historical research and archeology.[2] Their mission was to perform a historical and archaeological assessment of the eastern section of the Washington Valley Park. The study area straddled the first ridge of the Watchung Mountains and encompassed the land that had been identified on 18th century maps and long recognized as the site of Wayne's Brigade encampment in 1777. The area covered county-owned lands lying east of Middle Brook, south of Gilbride Road, west of Vosseller Avenue, and north of Route 22.

Hunter archeologists excavated, conducted metal detector surveys, and drew detailed maps of their findings. They then matched their results with land deeds and other archival documents. They also searched for credible evidence in the earlier studies conducted by A. A. Boom and William Liesenbein. Hunter's search was focused on finding evidence of a redoubt or fortification that guarded the pass that is now Vosseller Avenue, and also a signal tower which was reported to have been erected on the site. Archeologists prospecting The Hunter team found hundreds of military and civilian artifacts of the Revolutionary War period to confirm the existence of Wayne's encampment, but no conclusive evidence of a redoubt or signal tower was ever discovered. While busy defining the boundary of the encampment site, the Hunter staff identified four mid-18th century farmsteads along the mountain top. All were recognized by the foundation remains of houses, barns, outbuildings and a grid of farm lanes and stone walls. The Wayne Brigade encampment covers the site of the Oliver/ McCain/Campbell farmstead, but its precise boundaries may have extended north across Miller Lane and possibly across Vosseller Avenue.

The Hunter study recommended that the important Revolutionary War complex at Middlebrook be given better identification in the New Jersey and National Registers. They suggested that a thorough assessment of archaeological potential be undertaken to increase public awareness of the historical importance of the entire area. Hunter also recommended that high priority be given to the protection and public acquisition of archaeologically intact parcels of land in the general area of the Middlebrook Encampment where evidence of the military occupation still exists.

Written History is Limited

This book is the only dedicated book devoted entirely to the Middlebrook Campground that has ever been written in the two-and-a-half centuries since its occupation by the American Army. Recorded accounts of the camp are limited to a few academic papers, archaeological studies, and journal articles. It is astounding that so little has been written over the centuries about the encampments at Middlebrook and the adjacent areas encompassed by the site. The other large winter Revolutionary War cantonments of the war—Morristown, Valley Forge and New Windsor—have been commemorated with a plethora of books, articles and memorial events over the years.

Historian Benson Lossing provided the first documentation in

1851, a brief description of a small area within the site and the rock lookout he found there. The next written account appears in three pages of *The Centennial History of Somerset County*, written in 1878 by Reverend Abraham Messler, pastor of the First Church of Raritan. He visited Wayne's 1777 campground on the crest of the first ridge in 1837.3 The remnants of huts still remained there at the time. In 1912, the Reverend T. E. Davis of Bound Brook provided a concise ten-page summary for the *Somerset County Historical Quarterly* which described both occupations of the camp.[4]

More in-depth works began to appear forty years later. In 1952 Peter Angelakos, a student at Rutgers University, wrote an essay on Camp Middlebrook for the Society of Colonial Wars. This twenty-page summary relies almost entirely on a single source, *The Writings of Washington*.[5] The most comprehensive study of Middlebrook, *The American Eagles Nest*, was written by Carl E. Prince in 1959. This eighty-five page study, a thesis prepared for a master's degree, has stood as the primary source on the Middlebrook Encampment since that time. The pamphlet is professionally written and documented, but unfortunately it only covers the 1778-1779 winter encampment.[6]

Existing archeological studies are limited to specific locations and focus on the technical methodology of excavating, rather than history. The most significant of these works are the findings of Hunter Research. The writings of Max Schaubisch, Clifford Sekel, and John Seidel are limited to the Pluckemin Artillery Camp. These works are explored in the next chapter. Several autobiographies, biographies, letters, and papers of American and British military leaders who were at Middlebrook Encampments provide occasional primary source information.

Mapping Middlebrook

Fortunately, the entire Bound Brook area was extensively mapped by both American and British cartographers during the years of the

Revolutionary War. These documents include topography and place name identification. The most informative are the many military maps by Erskine and DeWitt, drawn between 1777 and 1781, and John Hills' from 1777 to 1782.

During the encampment General Washington urged Erskine to move into the camp. His maps and knowledge were so valuable that he risked being kidnapped by British intelligence. Washington also commented on the superb quality of his maps and indicated that he used them on many occasions.[7] The originals are held by the Library of Congress, The New York Historical Society and the William L. Clements Library at the University of Michigan. However, copies can be reviewed locally at Washington's Headquarters, Morristown Historical Park and the Alexander Library at Rutgers University in New Brunswick.[8]

PART IV

The Artillery Camp at Pluckemin

Chapter Twenty-Three

With Knox at the Artillery Camp

THE ARTILLERY BRIGADE, commanded by Brigadier General Henry Knox, functioned as an independent unit within the Continental Army. Knox established a separate camp at Pluckemin for his force of 700 to 800 men during the winter of 1778-1779. The small hamlet was seven miles west of the Middlebrook main army camp. The village was nestled on the western side of the second Watchung Mountain in what is now the southern section of Bedminster Township, New Jersey. It lies on higher ground, near the Jacobus Vanderveer House, on Route 202, the road from Middlebrook to Veal Town, now Bernardsville. A housing complex known as "The Hills" covers the area today.

Henry Knox, a former Boston bookseller with an interest in military history, was a twenty-five-year-old artillery officer in 1775. He had gained a reputation for military brilliance early in the war and had been appointed as an officer in the Continental Army. His first mission was to lay out defense positions around the city that was under siege by the British fleet. General Washington was appointed to lead the Continental Army in July 1775. When he first inspected the defences he was amazed by the expertise of Knox.

Knox and the Ticonderoga Cannons

Benedict Arnold, Ethan Allen, and the Green Mountain Boys of Vermont had captured Fort Ticonderoga in New York in May. The

In 2012, the Friends of the Jacobus Vanderveer House commissioned a painting of General Knox at the Pluckemin Artillery Barracks, 1779. It was produced by New Jersey artist John Phillip Osborne, and now hangs above the mantel in the house's 1813 Main Parlor.

large British bastion contained many cannons. Knox proposed a courageous plan to recover and transport the captured British cannons from Fort Ticonderoga to Boston. The fort was hundreds of miles away, near Lake Champlain. It would require hauling sixty tons of heavy guns, on ox-drawn snowsleds, over 300 miles of frozen rivers and ice-covered mountains. The daring feat required bringing the heavy cannons across a large lake, and over frozen rivers.

Washington agreed to send Knox to attempt to retrieve the cannons. On December 5 Knox reached snowbound Ticonderoga, where he selected fifty-eight cannons to move to Boston. The artillery included heavy mortars and howitzers. Several were twenty-four pounders known as "Big Berthas." They were eleven feet long and each weighed 5,000 pounds.

The cannons were carried to the northern end of Lake George and loaded onto a ship. The added weight caused the ship to run aground on a rock and it began to sink. A disaster was averted by pumping out the water. The voyage resumed and the cannons arrived safely at the southern tip of the lake. Storms had covered the craggy terrain with deep snow. Knox's brigade built forty-two sleds to pull the cannons

Knox moves cannons from Fort Ticonderoga to Boston. *(U.S. National Archives)*

across the wilderness using eighty yoke of oxen (a yoke is a wooden crosspiece that is fastened over the necks of two animals). Several cannons broke through the ice on the frozen rivers, but somehow were retrieved.

The procession was stalled by a two-foot snowstorm on Christmas Day. John Adams wrote that he saw the "noble train of artillery," as the equipment came to be called, pass through Framingham, Massachusetts on January 25. The weapons arrived at Cambridge, just outside Boston, on January 27, nearly two months after leaving Ticonderoga.

During the night of March 4, the Americans dragged the artillery up onto the Dorchester Heights, which overlooked both the city and the harbor. The next morning, the British were stunned by the sudden appearance of these forbidding weapons. General Howe is supposed to have said "My God, these fellows have done more work in one night than I could make my army do in three months!" Confronted by this menace, 9,000 British soldiers and their dependents, along with 1,000 Loyalists, evacuated the city on March 17, a day still commemorated as "Evacuation Day" in Boston. They sailed to New York City and never returned to northern New England.

Early in 1776, less than a year after the war started, Knox sent a memorandum to Congress recommending a series of actions that he believed were essential to achieving military success by the effective use of artillery. These measures included the establishment of an academy where the theory and practice of gunnery could be taught. Congress considered the proposal extravagant and did not approve. In September 1776 the ever-persistant General Knox again submitted his recommendations to Congress for upgrading the artillery and establishing a training facility:

Headquarters, Harlem Heights, September 27, 1776.
The following hints for the improvement of Artillery of the United States is humbly submitted to the Committee of the honorable Congress now in Camp.
~ That there be one or more capital laboratories erected at admittance from the seat of war in which shall be prepared large quantities of ordnance stores of every species and denomination.
~ That there be in the same place a sufficient number of able articifers be employed to make carriages for cannon, of all sorts and sizes, ammunition wagons, tumbrils, harness, etc.
~ That as contiguous as possible to this place a foundry for casting brass cannon, mortars, howitzers be established upon a large scale.
~ And as officers can never act with confidence until they are masters of their profession, an academy established on a liberal plan would be of utmost service to America. Where the whole theory and practice of fortification and gunnery should be taught, to be nearly on the same plan as that at Woolrich, making allowance for the difference in circumstances, a place to which our enemies are indebted for the superiority of their artillery all who have opposed them.
~ That these and other matters respecting the Artillery and Artillery Stores be under the direction of a board of ordnance whose business shall be the regulation and management of the affairs of this department, and to whom returns shall be made.
~ The Corps of Artillery now in the service of the United States

Henry Knox. (From *National Portrait Gallery of Eminent Americans... from Original Full Length Portraits by Alonzo Chappel*)

is exceedingly insufficient for the operations of an extended civil service. It consists of a little more than six hundred officers. Of these, one hundred are in the northern army where their numbers are unequal to the service.[1]

When Knox choose Pluckemin, New Jersey he must have envisioned it as the place where he could achieve his 1776 dreams. There he could establish an academy to train gunners and build a base to manufacture and repair heavy weaons. Later that year General Washington forwarded Knox's proposal to Congress, requesting "the establishment of Continental Artillery, magazines and laboratories."

The proposal was finally approved, and Knox prepared a blueprint for the first dedicated American military academy. This plan for an officer training facility was drawn twenty-four years before the founding of the United States Military Academy at West Point. The location of the academy at Pluckemin distinguishes the Middlebrook Encampment from all others of the Revolutionary War.

The Artillery Camp with its academy, so eulogized in American history for its military significance, had other less incredible negative attributes. An examination of regimental orderly books and other documents written at the camp reveal distressing events, including a mutiny and rampant desertions. Few historians are acquainted with the dark side of venerated Pluckemin. Also astonishing is that its location was lost for over 200 years.

Population at the Camp

The main American army, about 8,000 soldiers, was stationed in the hills above the town of Bound Brook, about seven miles away at Middlebrook. Normally artillery regiments were broken up and attached to infantry regiments, but this winter it would be different. When the artillery companies reached winter camp on December 17, 1778, they regrouped and joined together at Pluckemin.[2] This concentration of men and cannons required large, independent facilities.

The table of organization for the Continental Army in 1778 specified that regiments consisted of twelve companies of sixty officers and men, for a total of 1,320. However, most units never reached full strength during the war. The average company at Pluckemin had between thirty and forty men. With twenty-two companies, the estimated population at the base would have been about 700 to 800 men fit for duty. A count taken in May, when the camp began breaking up, shows only 507 men. However it is likely that some companies had already departed on new campaigns.

Research from Original Sources

Fortunately, the research sources for the Pluckemin Artillery Camp are far more comprehensive than those of the main Middlebrook Campground. Complete regimental orderly books have survived. The records for Colonel Lamb's Artillery Regiment provide an account of daily activities, as well as all division-, brigade- and regimental-level orders. The documents include details of court martial cases, detailed information about camp life, and announcements issued by the Continental Congress.

Orderly books are the best source for the study of the "microhistory" of the Continental Army. The books provide a great many of the "missing pieces" of the army's history, its operations, and life within the officer corps and among the common soldiers. Based on the dates of entries, a given regiment can literally be tracked on a day-by-day basis.[3] This extensive and comprehensive primary source is supplemented by the correspondence and papers of many of the high ranking officers associated with the campground. The papers of Generals Washington, Greene, Stirling, Wayne, Knox and others make frequent references to the activities at Pluckemin.

These primary sources are supplemented by more recently written accounts by three archeologists. Max Schrabisch, a German archeologist, discovered remnants of the artillery camp in 1916. The science of archeology was primitive at that time, and Schrabisch did not keep field records of the artifacts that he found, nor where they were located about the site. His extensive collection of found objects has been lost. Schrabisch described his activities in a series of newspaper columns he wrote in 1917 for *The Bernardsville News*, a local newspaper. He also summarized his work in an extensive article in the *Somerset County Historical Quarterly* in 1917.[4] Unfortunately, these writings are viewed by historians and archeologists as having little value.

Historian Clifford Sekel uncovered documentary references to the Pluckemin Camp and rediscovered the site in the 1960s. He described his work and its historical significance in his master's degree thesis in 1972. Sekel was joined in field work by Rutgers archeologist John Seidel in the late 1970s. Seidel's field work over the next seven years provides the major source of much that is known about Pluckemin. These comprehensive academic studies cover both archeological work and the historical background of the site. Many of their findings are confirmed by a remarkable contemporary drawing discovered in the 1979 which depicts the buildings at the camp—"A South-West Perspective View of the Artillery Barracks, Pluckemin, N. Jersey," signed by I. Lillie, an artillery captain at the camp.

The Artillery Corps Goes to Its Own Campsite

Knox's Artillery Corps left Fredericksburg, Westchester County, New York on December 1, 1778 for the winter encampment at Middlebrook. Their march to Middlebrook with the Pennsylvania battalions was interrupted for the same reason that the main army also halted enroute to New Jersey. On December 4 a British fleet had unexpectedly sailed up the Hudson River as far as Tarrytown.

This incursion of fifty-two ships with a large landing force was planned as an attempt to take back the thousands of British troops captured at Saratoga who were being marched to captivity in Virginia. What made this threat terrifying was that the British could attack the Continental Army when it was most vulnerable while crossing the Hudson River between Verplancks Point and Kings Ferry. The strung out American columns could also be intercepted as they moved south through New Jersey to their winter camp at Middlebrook. A successful attack would cut the American forces in half, and the Patriot base at West Point could potentially be captured. Fortunately, the British forces had confusing orders. They arrived late and their plan miscarried. When the anticipated attack did not occur, the Continentals resumed their march south after a two-day delay. The artillery brigade, with hundreds of cannons mounted on horse-drawn carts, called limbers and caissons, began arriving at Pluckemin in early December 1778.

An Elite Corps

American cannoneers worked and drilled according to their own regulations and had little in common with the regular infantry regiments. The artillery brigade was a separate branch of the army made up of three battalions. They were not sponsored by any specific states, and were not included in the regular army chain of command. General Henry Knox, the commander of the artillery, reported directly to General Washington. Of the three battalions in the brigade two were from Pennsylvania and one was from Viriginia. The Pennsylvanians were commanded by Colonel John Lamb and Lieutenant Colonel Ebeneser Stevens. Lieutenant Colonel Thomas Forest led the Virginia battalion.

Artillerymen considered themselves to be an elite corps and often regarded the regular infantry troops as inferiors.There was little fraternization between the units. General Washington encouraged this autonomy by allowing Knox to lead independently and administer his own rules. Since he could select a location that would exactly meet his requirements, Knox saw no reason for the artillery to camp to be with the main army at Middlebrook for the winter, so he set up his own winter quarters seven miles away.

Artillery soldiers even had a special name. An artilleryman was called a matross. The word was derived from a German term meaning "sailor." The tasks of firing, loading, sponging, and handling guns were considered to be much like a sailor's work. The skills of the matross were highly valued. Few soldiers had the skill to compute the simple geometric calculations necessary to place a cannon ball on target. This ability was considered an art, and the intricate procedure for firing and commanding a gun crew made the artillery officers essential to a successful campaign. Matrosses were armed with muskets and bayonets, as their duties included guarding the guns and wagons on the march and assisting when breakdowns occurred. They also prevented the civilian teamsters, often under contact to transport the guns, from running away when the fighting began.

Artillery officers were often the best educated and trained leaders in the art of war, and had a reputation for being arrogant. Older men were often assigned to the artillery, perhaps because the duties were physically less demanding. They rode on the limbers and caissons while the infanty marched on foot.

The Guns

Cannons were the kings of the battlefield and the weapons of mass destruction of the Revolutionary War. Foot soldiers who were not supported by artillery were usually defeated. The artillery controlled the outcome of battles by besieging fortifications or bombarding a battlefield.

Cannons used gunpowder to shoot a round projectile. These weapons were mounted on sturdy wooden carriages designed to support the weight of the gun and its support structure. Cannons were classified by the weight of the ball they fired. For example, a "three pounder" cannon shot a three pound ball.

The effectiverange ofRevolutionary War cannon varied wildly, but typically it did not exceed 1,000 yards. The most commonly usedmobile cannonsduring the Revolutionary War were the three-pound "galloper" and the steadier six-poundfield guns, although largercannonup to eighteen pounds were used in some engagments. The smallest standard cannon was the two-pounder. The largest were fifty-pounder stationary cannon mounted in forts or on ships.

The American artillery train was a diverse assortment of guns.

In 1778 Knox considered this list to be the desired armament for a fully equiped regiment:

> Brigade artillery, 17 brigades, with four guns each: - 68 pieces to be 3, 4, or 6 pounders; with the park - two 24-pounders, four 12-pounders, four 8-inch howitzers, eight 5 1/2 -inch howitzers, ten 3 or 4 pounders, ten 6-pounders, for reserve, to be kept at a proper distance from the camp, thirty 3, 4, and 6-pounders, two 12-pounders, one 24-pounder, all the foregoing brigade, park , and reserve guns and howitzers to be brass. In addition, twelve 18-pounders, twelve 12-pounders, battering pieces, on traveling carriages, together with two 5 1/2 -inch and twelve 8, 9, and 10 inch mortars; the battering pieces and mortars to be cast in iron.

The effectiveness of artillery in the Revoutionary War was limited by two factors: the difficulty of mobility, and the time it took to load and fire each weapon. These guns were so unwieldly that officers on both sides had to carefully place their artillery in position before a battle to achieve accuracy. Errors in positioning could render the weapons useless during a battle. Because these weapons were so difficult to move, artillery would often change hands throughout a battle as infantry advanced or fell back. Captured artillery was seen as something of an valued prize of war. Both sides routinely lost and gained artillery pieces during an engagement.

Selecting a Site: Why Pluckemin?

Richard Frothingham, a deputy commissary of military stores, was sent ahead in late November 1778 to locate a site for armorers to work, and where a large amount of supplies could be stored. The place he selected was in the northern part of what would become the future artillery camp. His selection, Pluckemin, is said to have been named for a village in Scotland.[5]

It was an ideal site for the entire Knox brigade to spend the winter. Frothingham began construction of an armorer's shop, while at the same time gathering lumber for additional building projects. The place had an abundant supply of timber for building, heating, cooking, and for making charcoal for the armorers forge. The area was strewn with rocks, which could be used for building chimneys and foundations. Nearby lime pits could produce mortar. Streams at both ends of the camp provided an uncontaminated source of water for all needs.

Archeologist Clifford Sekel explored the topography and measured the site. He described it as a flat plateau at the western foot of the second Watchung ridge. The plateau is about 500 yards long and between 125 and 175 yards wide. A mountain rises sharply to about 375 feet above the plateau. From the western edge of the flat area the ground slopes down for a quarter of a mile to the road from Pluckemin to Bedminster, now Interstate 287. At both ends of the campsite are ravines which cut into the mountain.[6]

The Pluckemin site had the same advantage as the main encampment at Middlebrook. It was located at an important 18th century crossroads on a direct route to Morristown, fifteen miles away. It was in the rear of the main army at Middlebrook, which offered protection and safety from a surprise enemy attack. Today, it is near the junction of Routes 202 and 206 and Interstates 78 and 287.

The site had natural defenses. It was elevated above the surrounding area and a mountain rose sharply behind it. In locations north and south of the camp streams had cut deep ravines into the sides of the surrounding hills. To the west and northwest was the North Branch of the Raritan River. The local roads at the time were reported to be durable enough to support guns weighing up to three tons.[7] While the roads were ideal for transportation, they also provided avenues for an enemy attack. The camp was also exposed by long stretch of open fields and could be observed by opponents from over a mile away. However, a frontal assault could result in heavy losses for the attackers as they would have to advance without cover over this open terrain.

Camp Layout

The Pennsylvania Packet described the new Artillery Camp on March 6, 1779:

> *The huts of this corps are situated on a rising ground, at a very small distance from the road, and unfold themselves in a very pretty manner as you approach. A wide range of field pieces, Howitzers and heavy cannon, make the front line of a parallelogram and the other sides composed of huts for officers and privates. There is also an academy where lectures are read on tactics and gunnery and work huts for those employed in the laboratory. This military village is superior in some respects, to most I have seen.*

Along the north side side was a laboratory, an armorers shop, an artificer's workshop, storage facilities, and a park for artillery wagons The side at the base of the mountain was the first line of barracks and a line of storage huts. Bake houses, mess rooms and officer's huts were set into the mountain behind the huts. The cannons were lined up along the southern side and trained over flatland to defend the post.

The Pluckemin Encampment was the artillery portion of the seasonal cantonment of the Continental Army during the winter of 1778-1779. Its site was well protected by the presence of the main army along the first ridge and its seven mile distance behind the front lines in the Raritan River Valley. Despite the short duration of occupation, chronic supply problems, and discipline troubles the accomplishments of General Knox there affected the course of the remaining four years of the war. The establishment of the academy with an industrial center was the first successful attempt by the army to combine the two functions. The effective use of artillery refined at Pluckemin led ultimately to American victory.

Chapter Twenty-Four

A Well-Built Surprise

An Academy and a Beacon

HENRY KNOX ACHIEVED fame after his astonishing feat of delivering the Fort Ticonderoga cannons to Boston. Shortly thererafter he befriended General George Washington and quickly rose to become the chief artillery officer of the Continental Army.

As soon as he arrived at Pluckemin, he ordered that a large and substantial structure be constructed to house an academy. But first, temporary shelters, a tent camp, and ragtag huts were set up near the construction site. The initial priority was to build permanent barracks to house the soldiers before the weather grew colder. Work commenced on a line of barracks to hold about 250 soldiers, and they were finished by February. A large building, which eventually became the academy, was started in early December and had been completed by January 5, 1779. It was large enough to house fifteen companies, roughly four hundred men.

Barracks Construction

Skilled craftsmen and others working on the construction were divided into parties of fifteen men, all under the direction of one officer. The orderly books specified the work rules: "They are to begin at troop beating [7:00 a.m.] and work till one o'clock when they are to be dismissed one hour for dinner. Then begin and work half an hour after four o'clock. This is routine is to be continued until barracks are completed."[1]

Skilled workers were assigned to more intricate jobs. These included framing and laying out dimensions from the construction plans. Men without specialized skills were employed in felling trees, clearing and leveling building sites, and hauling materials. The unique barrack construction required heavy logs and timber beams which were hauled by teams of horses and lifted by hand into position. Artillery soldiers used their special expertise acquired by handling massive guns and dragging weighty caissons, the wagons that carried the cannons.

As the end of the year approached the weather grew colder, and General Knox had difficulty motivating the freezing and often hungry men. The truculent gunners knew they had unique and valued military skills and resented being employed as common laborers. But Knox insisted that the barracks be completed on schedule. He increased the daily construction crew to 200 men and more men were assigned to cut shingles.[2]

Food became scarce in the camp and a serious shortage of shoes added to the men's misery. Aside from the lack of meat and bread, Knox also complained that small necessities were lacking. "Indeed, we have a scarcity of little necessaries such as sugar, chocolate, etc."[3] The desperate Knox authorized two gills of rum to be issued daily to each man to induce them to work. A gill was equal to a quarter of a pint. A severe snowstorm the day after Christmas added to the ongoing agony.

Knox, although usually compassionate, was relentless in getting the buildings completed, and he threatened to punish those who shirked. He pleaded, "The necessity of having the barracks completed is urgent and calls for the greatest exertions."[4] Despite the general's demands, the building construction was not finished until the middle of January in 1779. Knox bragged to General McDougall on January 10 that his men had "got into their barracks which are comfortable and on an elegant plan."[5]

General Washington and all senior officers at Middlebrook were amazed when they first visited Pluckemin and witnessed what the

A Well-Built Surprise

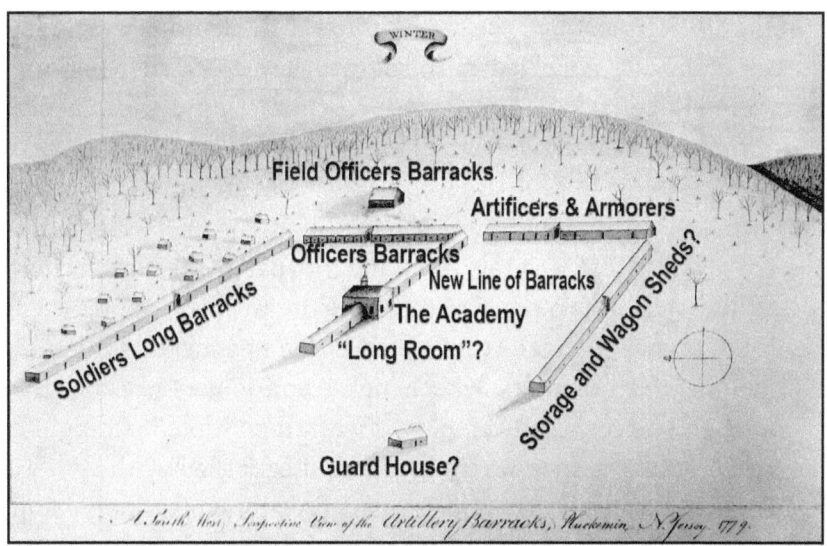

1770s drawing of the Pluckemin Cantonment from Captain John Lillie, with modern overlay. *(Morristown National Historical Park)*

artillery men had accomplished. They had not built standard huts, but had constructed much larger and more durable structures. The buildings were spacious and built of boards, not logs, and had fireplaces and chimneys.[6]

The Essential Role of Artificers

Craftsman in the artillery brigade who had the special skills needed to move and repair cannons and other weapons were called artificers. They were the expert artisans and mechanics who kept military equipment in good working order. Artillery artificers were officially recognized as an occupational speciality in 1779 at Middlebrook. The muster roles at Pluckemin show that fifty-six officers and men assigned to the artillery brigade were designated as artificers. Their areas of expertise included wheelwrighting, gunsmithing, blacksmithing, and harness making.

The artificers shop for the main camp at Middlebrook was under the command of Colonel Jeduthan Baldwin. It was located in Raritan, near Somerville. Blacksmiths, wheelwrights, tailors, shoemakers, and other craftsmen worked there repairing small arms and mending clothing. In February 1779 Baldwin's regiment was assigned 186 artificers.

These skilled craftsman also took an active part in the construction of the barracks and other structures at the artillery camp. Their participation was mandatory, as stated in the Regimental Orders at Pluckemin: "All carpenters, wheelwrights and joiners will join the artificers and be constantly employed with them while building the barracks."[7] A special shop was built on the site for their work.

The complexity and superiority of many features of artillery camp construction were evidence of the contributions of the artificers. They provided the skills that enabled the buildings to be completed with more precision than that found anywhere at Middlebrook. This pool of talented workers was also capable of instructing other men to perform more proficient work. While excavating the ruins at Pluckemin, archeologist Clifford Sekel reported:

> The buildings to house the men at Pluckemin were built on raised, dry stone foundations and probably contained some form of wooden flooring. The abundance of stone at the campsite allowed the men to build fireplaces and chimneys of stone and mortar. The armorers shop was a ready source of nails for roofing and flooring as well as other ironwork. The extra tools for this work came from either the company's equipment or in military stores at the main camp.

The shivering soldiers finally moved into the barracks on February 3, 1779. Officer's quarters, guard houses, an armorer shop, a forge and storage facilities were also completed at that time. Facilities for the reclamation of gunpowder and ammunition, a powder magazine, a mess room for dining, and baking facilities were also completed.

The Academy - America's First West Point

Knox envisioned an academy for officer training with the plans he sent to Congress two years before the Pluckemin encampment. He was so obsessed with the need for instruction that he initially started a school in a large marque tent when his brigade marched to New Jersey. His proposals for an institute were officially approved when he arrived. Soon a laboratory was installed and the shop was set up for artificers, and organized training classes began.

General Knox enlisted Christopher Colles, an engineer and inventor, as" preceptor" of the academy. The academy focused on the training of artillery and engineering officers. Knox explained the purpose of the institute:

> *General Knox states the Academy is to be opened on Monday next when Mr. Colles the preceptor will attend every day in the week Sunday excepted for the purpose of teaching the Mathematicks & cc. ...As the Officers of the Corps will be those means have an opportunity of acquiring a more particular and expansive knowledge of the profession and making themselves better qualified to discharge the duties of their respective stations — The General expects that they will apply themselves in good earnest to the study of this so essential & necessary Branch of Science — The duty they owe themselves — a regard for their own reputation and the just expectations of their Country: The General hopes will induce every Officer to pay the closest & most diligent attention.*[8]

The academy had a large central building. It was erected on a narrow terrace forty feet above the camp plateau and was the most imposing structure of the campground. The main room of the building was thirty by fifty feet. It had plastered walls and an arched ceiling with a cupola on top. At the southern end of the main room there was a long room, twenty feet wide by 400 feet long.[9] A regimental orderly book described the building: "The Academy was raised several feet above the other buildings and capped with a small cupola, which had a very good effect. The great room was 50 feet by 30, arched in an agreeable manner and neatly plastered within."[10]

The academy building was the hub of Pluckemin, and all other structures of the encampment were built around it. Important activities and gatherings of all kinds were held there. Few officers attended church there even though religious services were held each Sunday. When Knox threatened them with discipline for not attending services his warnings were ignored. The general did not follow-up his admonishments.[11] Despite Knox's enthusiasm, attendance at the academy training classes was disappointing during the entire encampment. The problem was lack of space. Lecture rooms were in constant use for court martial hearings, social events for officers, and church services.

The Greatest Social Event

A grand ball was held at the academy on February 18, 1779 to celebrate the anniversary of the alliance with France. Surgeon Thacher reported on the attendees: "General Washington and his lady, principle officers of the Army and their ladies and a considerable number of respectable ladies and gentlemen of the state of New Jersey."[12]

Before this occasion General Knox issued a dress code for the soldiers of the camp. "The men are to be clean dressed, their hats cocked, coats hooked back and their hair braided and turned up behind and powered. The General desires to see that none of the men are absent on that day and these orders respecting their appearance are complied with in their fullest latitude."[13]

This was the greatest social event at the Middlebrook Camp during that winter of the second encampment. Guests from all over New Jersey started arriving in carriages and were paraded down a 100-foot-long pavilion. Thirteen huge paintings were on display which depicted colorful scenes of the war. Sixteen cannon blasts invited guests to a multi-course dinner which was followed by a fireworks display and an evening of dancing that lasted until dawn. The proud Knox reported, "We had about 70 Ladies, all of the first run [most attractive] in the state and between three and 400 gentlemen." General Washington led off the party by dancing with Knox's wife Lucy.

The Pennsylvania Packet or the General Advertiser (Philadelphia) for March 6 published a letter "from a foreigner to a gentleman in this city," that vividly described the entertainment:

Near Middlebrook Camp February 22

His Excellency the Commander in Chief arrived from his Headquarters about three o'clock in the afternoon. Mrs. Washington was in a carriage, accompanied by that steady friend to the rights of mankind, Mr. [Henry] Laurens, the late President of Congress. I had also the pleasure of seeing Mr. [William] Duer, late a Member of that Honourable Body from the state of New-York. I was introduced to Mrs. Washington, Mrs. Greene, Mrs. Knox, and a circle of brilliants, the least of which seemed more valuable than that stone of immense price which the King of Portugal received from his Brazilian possessions.

About four o'clock the occasion was announced by a discharge of thirteen round of cannon. We then repaired to the academy to dinner. The company was composed of the most respectable Gentlemen and Ladies for a considerable circuit round the camp, and as many of the officers of the army as could possibly attend.

I had, till now, only seen the outside of the academy. It was raised several feet above the other buildings, and capped with a small cupola, which had a very good effect. The great room was fifty feet by thirty, arched in an agreeable manner, and neatly plaistered within. At the lower end of the room was a small enclosure, elevated above the company, where the preceptor to the park gave his military lessons. This was converted into an

Dancing at the Greene's. *(U.S. National Archives)*

orchestra, where the music of the army entertained the company. The stile of the dinner was of that happy kind, between the extremes of parade and unmeaning profusion, and a too great sparingness and simplicity of dishes. Its luxury could not have displeased a republican. The toasts were descriptive of the day, while the joy and complacency of the company could have given umbrage to none, except our enemies the British.

Just as night came on, we were called upon to the exhibition of fire works. These were under the direction of Colonel Stevens of the artillery. The eye was very agreeably struck with the frontispiece of a temple, about one hundred feet in length. It was divided into thirteen arches, each arch embellished with an illuminated painting, allegoric of the progress of your empire, or the wise policy of your alliance; the center arch was ornamented with a pediment, and proportionably larger than the others; the whole supported by a colonade of the Corinthian order. The different works in pyrotechny were very agreeably disposed, and displayed to great advantage.

In all public rejoicings, I make it a point to mix with the multitude; if they are not pleased, the demonstration may be considered as wrong. In the present instance I was charmed to find that every man's heart went along with the occasion.

When the fire works were finished, the company returned to

the academy, the same room that had served to dine in served to dance in; the tables were removed, and had left a range for about thirty couple, to foot it to no indifferent measure. As it was a festival given by men who had not enriched themselves by the war, the lights were cheap and of their own manufacture; the seats the work of their own artizans; and for knights of different orders, there were hardy soldiers, happy in the thought of having some hand in bringing round what they were celebrating.

The ball was opened by his Excellency the General. When this man unbends from his station, and its weighty functions, he is even then like a philosopher who mixes with the amusements of the world, that he may teach it what is right, or turn its trifles into instruction.

As it is too late in the day for me to follow the windings of a fiddle, I contented myself with the conversation of some one or other of the Ladies during the interval of dancing.... I do not recollect that I have ever been more pleased on any occasion or in so large a company: There could not be less than sixty Ladies.

The Academy Vanishes

It is unknown how long the academy building lasted after Knox and the artillery left Pluckemin in June 1779. Like the barracks, it was used for several months as a hospital facility. All traces of the academy building vanished, along with other evidence of the encampment, soon after the end of the Revolutionary War.

Years later, when Knox became Secretary of War in 1794, he met with President Washington and Secretary of the Treasury Alexander Hamilton to propose establishing a permanent military college for the new nation at West Point. In creating the new school he relied on the experience at Pluckemin. The place has always been considered the predecessor of West Point, the military academy established twenty-three years later. It is unknown if there were other efforts to establish

Replica model of a beacon. Not to scale. *(Summit Historical Society's Carter House museum)*

military schools during the interim before West Point opened. Although the academy at Pluckemin functioned for only six months, it is recognized as America's first military academy and is listed as such on the National Register of Historic Places.

A Beacon with a Warning Network

Since the high ground at Pluckemin had a panoramic view of the surrounding area, it was selected as one of the sites for an alarm signal beacon. This beacon could alert the local militia in case of a surprise attack by Henry Clinton's forces out of New York. In March 1779, General Washington wrote to New Jersey's Governor Livingston, "The possibility of the Enemy's making incursions into the state has suggested the expedience of fixing signals to communicate the most speedy alarm to the country; and of appointing convenient posts for the Militia to assemble to orders."

That same day Washington ordered General Stirling to build a series of signal beacons "on conspicuous hills and mountains, which appear to be judicious and well disposed." All beacons would be placed on the eastern side of the Watchung Mountains in New Jersey. Stirling, a gentleman farmer and soldier from nearby Basking Ridge, was familiar with the local topography. The Watchung Mountains were a natural fortress extending over forty miles from Somerset County to Bergen County. With signal beacons, the hills could became an impregnable line of defense.[15] Twenty-three beacons were eventually built. (See Appendix G for the specific locations of each beacon).

On March 23 Washington wrote to General Knox, ordering him to erect one of the signals on the mountain to the rear of Pluckemin:

> *For the more speedy assembling of the Militia upon and emergency, I have agreed with the Field Officers in this and the next count to erect Beacons on the most conspicuous hills, the firing of which shall be signals for them to repair to their different Alarm Posts— You will be pleased to have one erected upon the Mountain in the rear of Pluckemin, upon the place that shall seem the most visible from the adjacent County. The Beacons are proposed to [be] built of logs in the form of a pyramid 16 or 10 feet square at the Base, and about 20 feet in height, the inner part to be filled with brush— Should there be occasion to fire it, you will have proper notice.*

After the barracks were completed, a signal beacon was built on the mountain in the rear of the artillery camp. As directed by Washington, it was filled with dry brush that could be easily ignited in the event of an attack. The alarm to set the signals fire blazing was three cannon blasts. The fire could be seen from the main camp at Middlebrook.[16] Other signal towers were built nearby at Steele's Gap (Foothill Road, Bridgewater), Wayne's Gap (Vossellor Avenue, Martinsville) and Quibbletown Gap (Warren Township).

The beacon system was never put into use during the Middlebrook Encampment. However, the next year the beacons proved their worth. Two British invasions failed after beacon fires alerted the New Jersey Militia.[17] They were able to mobilize, quickly move to the threatened areas, and block British attempts to attack the Continental Army at its encampment in Morristown at Jockey Hollow. Largely because of the beacon system, the British were never able to dislodge Washington from his winter camps or penetrate his defenses. The New Jersey signal system was the most sophisticated early warning method built during the Revolutionary War.

Chapter Twenty-Five

Our Compassionate General

Headquarters at the Vanderveer House

HENRY KNOX HAD rudimentary artillery training with the colonial militia but most of his knowledge was gained from reading military books in his Boston bookstore. Somelessons were learned the hard way. He managed to blow off two of his own fingers while firing a fowling musket at wild birds. In 1770, Knox was a witness to the Boston Massacre. According to his affidavit, he attempted to defuse the situation by trying to convince the British soldiers to return to their quarters.

Before the war, Knox was an enthusiastic Patriot and an active member of the zealously nationalistic society known as the Sons of Liberty. This group of instigators and provocateurs used civil disobedience, threats, and in some cases actual violence to intimidate loyalists and to outrage the British government. Their goal was to push moderate colonial leaders into a confrontation with the Crown.

Knox and his wife Lucy slipped out of Boston for a safer place when the war broke out with theaction at Lexington and Concordon April 19, 1775. His abandoned bookshop was looted and all of its stock was destroyed or stolen. He rejoined the militia and used his military engineering knowledge, gleaned for the most part from reading, to build fortifications around the city. While serving under GeneralArtemas Ward during theBattle of Bunker Hill he directed the defensive American cannon fire.

When General Washington arrived in Cambridge in July 1775 to take command of the army he was amazed by the work Knox had done on the fortifications. The two men soon became good friends. Knox did not have a commission in the fledgling Continental Army but his frequent interaction with Washington and other generals enabled John Adams to get a colonel's commission for him at theSecond Continental Congress.

Knox and his wife Lucy dined with the new commander-in-chief on several occasions, and GeneralWashington became fascinated with the young man's conception of the role of the army's artillery. On November 17, 1775, Washington appointed Knox as the commander of artillery for the entire Continental Army. A few weeks later, Colonel Knox set out on the perilous mission to Fort Ticonderoga to retrieve the artillery that the British had captured there.

The admirable and technically astute Knox had a shortcoming, his leniency in disciplining his troops. He affectionately referred to his men as "my poor rebels." He rarely punished them for even serious offenses and when he did, he usually forgave the offenders before sentences were carried out. His compassion was anticipated and often resulted in his orders being ignored, and a continuation of infractions.

Knox continually warned soldiers about minor offenses, such as cutting down trees for firewood near the camp or littering the campground with garbage. His orders were simply ignored. The colonel's slack approach to discipline soon caused a loss of control over the men in his brigade and the authority of his officers began to be challenged. Much of the blame for the deterioration of order at Pluckemin can be attributed to Knox's imprudent policy of allowing his officers to live away from the camp. Without direct supervision the men often found ways to get into trouble.

While the soldiers were engaged on construction projects during December 1778, there was little that could distract them from their strenuous nonstop activity. As a result, few restrictions regarding personal conduct were needed to maintain discipline. But after the construction of main camp structures was completed and the men were comfortably housed in the barracks, their days began to be filled with less physically demanding activities. Military routine and abundant leisure time led to boredom, and it became necessary to issue several new rules to maintain discipline in the camp.

Our Benevolent General

When Lamb's Regiment moved into camp on December 16, 1778, the population swelled, and problems began. Knox was informed that liqueur was being sold to soldiers from huts behind the barracks. He ordered the brigade quartermaster to investigate this activity, and to destroy any spirits that might be discovered, as well as the huts themselves. Liqueur was found, but there is no record of it being confiscated or any huts being destroyed.[1]

On February 9 a court martial was held. A soldier was convicted of drunkenness and was dismissed from the army. That same court convicted a bombardier and a matross for desertion. The court demoted the bombardier but sentenced the matross to be "picketed for ten minutes." Picketing required the offender to stand barefoot on a sharpened stake. Knox intervened, and prevented the punishment from being carried out. He also forbade the cruel penalty to be applied anywhere in the camp in the future. This pattern of leniency seemed to have the effect of encouraging others to commit petty crimes and to desert. These types of notices naming deserters from Pluckemin appeared in local newspapers soon after:

> *Deserted from Capt. Doughty's company, Col Lamb's regiment of artillery, Michael Rowe: he is about 6 feet high, dark complection, short black hair, grey eyes. Full faced about 20 years old; had on when he went off a short jacket and overalls.*
>
> *Bezeliel Ackly; he is about 25 years old, brown complexion, light brown hair, grey eyes; had on when he deserted a black regimental coat, hat and a pair of overalls. Any person that shall apprehend said deserters, and deliver them to the commanding officer of the regiment, at the park of artillery, shall be entitled*

to a reward of twenty dollars for each, with necessary expense attending the same. Thomas Thompson, Capt. Lieut. Art.

Deserted from the Park of Artillery this morning, William Helnit, matrross in Capt. Mott's company, Col.bs regiment of artillery; he is about five feet five inches high, stout built, fair complexion, light hair and blue eyes, blooming cheek , between twenty-two and twenty-six years of age, his hair is short and curls; was whipped the 14th instant for theft, which his back now shows sufficient proof: he is suspected as having stole from a man in the neighborhood of the Park last night, about two thousand dollars: of course he has plenty of money.

In February a court convicted six men of mutiny. The proscribed penalty for mutiny was death. Five of the men were sentenced to 100 lashes and a ringleader was sentenced to death by hanging. Knox used the opportunity to intimidate others who might be contemplating an uprising, and announced a new stern disciplinary plan to curb similar offenses. Each evening all troops would be called into formation at Tattoo (8 p.m.). The men would then be confined to their barracks or the campground until morning. Any soldiers found outside the camp after the curfew hour were to be arrested.[2] Knox confirmed all of these sentences and ordered preparations for the punishments to begin.

All troops at the camp were assembled on the parade ground to witness the floggings and execution of the six mutineers. But at the very last moment the compassionate Knox ordered the lashings not to be carried out and then fully pardoned the man who had the death sentence. All six men were sent back to duty with only a warning that any future offenses by them would be punished with the "most exemplary severity."[3] There are no details describing the nature of their mutinous offenses. It is likely that the men were either refusing to obey simple orders or were being insolent and disrespectful to officers. Unrest at the camp and a recent mutiny at Danbury had apparently exacerbated sensitivity and caution among all commanders at Middlebrook.

The Pluckemin Mutiny

The casual attitude of Knox toward discipline and loss of control that resulted finally triggered a mutiny. In February 1779 most soldiers at Pluckemin refused to work or follow orders. When the men began to threaten their leaders, sergeants and officers armed themselves and prepared to use force. Two ringleaders were tried the next day and both were found guilty. One was sentenced to die by firing squad and the other to receive 100 lashes.

Pennsylvania Line Mutiny, woodcut created 1881. *(New York Public Library)*

As was expected, the kindhearted Knox forgave both offenders. A few days later he followed up by then issuing another futile warning to the camp. "Mutiny is a crime of so aggravated and dangerous in nature as to place the criminal almost beyond the reach of pardon and should there be any more instances of this, the offender shall be punished with the most exemplary severity."[4]

The number of offenses in the camp steadily increased and Knox contritely acknowledged that his efforts to enforce discipline with leniency were not successful. On May 13, 1779, he ordered all officers and men in camp to witness the execution of three deserters, privates Robertson, Baker and Ford. As expected, the executions were soon cancelled.

On May 31 a contrite Knox finally admitted that his clemency was a mistake: "...of late there have been many more men confined for a violation of duty than has heretofore been known in the Corps and in some instances where I have remitted punishment it has not

Our Compassionate General 291

produced the good effects that was in hopes it would." Discipline at Pluckemin continued to deteriorate after the mutiny and the camp remained virtually out of control until the artillery brigade left Pluckemin in June.

The Role of Knox's Headquarters at the Vanderveer House

The Jacobus Vanderveer House served as the headquarters for General Henry Knox when the Continental Army artillery was located in the village of Pluckemin during the Second Middlebrook Encampment. The general lived there with his family from December 7, 1778 to June 3, 1779. Today, the house is located on the southbound side of Route 202/206, in Bedminster Township on the grounds of the 218-acre River Road Park. This house dates from 1772 and is typical of homes built by prosperous Dutch settlers in New Jersey. The meticulously restored Vanderveer House is the only building associated with the Pluckemin Artillery Academy that has survived. It is also the last structure that remains from the earliest settlement of Bedminster. This house is the only place for visitors to learn in detail the history of the Pluckemin Cantonment site.

Jacobus Vanderveer, who built the small farmhouse, was a descendant of one of the area's many Dutch families. He also owned a large farm on the north branch of the Raritan River. His father was a founder of the Bedminster Dutch Reformed Church, organized in 1758. He donated the land for the church building and the nearby cemetery. The church, once at the heart of the little village, was demolished, but the cemetery still lies just north of the house.[5]

General Knox described his quarters as being "comfortable and clever." He rarely left the home to attend social functions at the main Middlebrook camp. He was joined in the home by his wife Lucy and their two-year-old daughter, also named Lucy. During their stay, another daughter, Julia, was born. Sadly, this child died in infancy. Knox seemed to be oblivious to the chaos that reigned at the encampment only a few hundred yards away. He continued to occupy the comfortable farmhouse through the summer of 1779 when Washington ordered the final remnants of the army to move north to the Morristown area.

Following its occupation by the Knox family the home underwent a series of alterations, but it remained in the Vanderveer family until it was purchased at auction in 1875 by Henry Ludlow. The house was subsequently owned by the Ballantine and Schley Families, who utilized the property for hunting and polo. In 1989, the then neglected and decaying building and property were purchased by Bedminster Township, with the help of Green Acres funding.

Restoration began when the home was stripped of many alterations and largely returned to its original appearance. Much of the original house remains. The kitchen wing was rebuilt based on archeological evidence and documented data, and the kitchen hearth and the Vanderveer parlor have been restored. The flooring is made of wide pine boards. The house was registered in 1995 on the National List of Historic Places.

In 1998, a volunteer group, the Friends of the Jacobus Vanderveer House, was formed to restore, develop, and operate the house as a nationally significant historic site and museum. They helped restore the home, established historic collections, and supported research to tell the story of General Henry Knox and the Pluckemin artillery encampment. The house serves as a museum with exhibits interpreting the occupancy of Knox, the adjacent Artillery Park, the Vanderveer family and the Dutch-American culture of the Raritan Valley.

Today visitors to the house will see fully finished rooms and exhibits. A history center houses a collection of books and documents that chronicle the origin and restorations of the house, colonial architecture, and colonial life. The home that housed the Knox family provides a special insight into their personal activities during the days of the artillery camp. The house has a story to tell and uncovers many details of the lives of its private owners, social history, political affiliations and culture of rural Somerset County.

Chapter Twenty-Six

Everyday Adventures at Pluckemin

EVERYDAY ACTIVITIES at the artillery camp were repetitive and regimented. They followed the regimen proscribed in *Regulations for the Order and Discipline of the Troops of the United States,* the rule book written by Baron von Steuben during the previous winter at Valley Forge.

Von Steuben created a standard method of drills for the entire army. He wrote the book in French since he was not literate in English. Afterward, his instructions were translated into English, and copies were distributed to each company. Von Steuben then established a "model company" of 100 men for each brigade. In addition, fifty Virginians came from Washington's Life Guard to demonstrate the new exercises to the rest of the army. The drills were quick and simple, and von Steuben regularly worked himself with the troops. He taught the essentials of military tactics and discipline, and included administrative practices such as proper bookkeeping and how to maintain hygiene standards.

The American soldiers appreciated von Steuben's willingness to work directly with them. They were amused by his use of colorful words and expletives in several different languages and his reliance on an aide to curse at them in English. Von Steuben's guide, better known as the "Blue Book," was based on the model drill companies that he had previously formed and commanded. The booklet served as the country's official military regulations throughout the Continental Army. Many of its guidelines are still followed to this day.

The baron relentlessly drilled the troops at Middlebrook. He was essentially the U.S. Army's first drill sergeant. The troops emerged from the camp far stronger and better prepared for war than ever before. Their surprise victory at Stony Point, New York, only two weeks after they left the camp, was credited to von Steuben's techniques.

Each day the men drilled, stood guard, and were assigned special tasks. These included storing supplies, digging latrines, repairing buildings, and filling musket cartridges. About fifty men at Pluckemin were assigned daily guard duty. Officers supervised construction projects and attended classes at the academy. Life for the men during the construction of the camp filled their day with exhausting work. Few rules were needed to maintain discipline. After the structures were completed and the weather deteriorated there was a little to do. This ample excess leisure time led to frequent infractions and it then became necessary to impose many onerous restrictions at the artillery park.

Cleanliness

While he was considered to be imprudently lenient on most matters, General Knox constantly insisted that the troops maintain an impeccable personal appearance. The men generally complied with this direction, especially since many of them were apparently proud of their personal appearance. Soldiers were ordered to "come on parade in the morning with their hats cocked, shoes blackened, faces shaved and hair combed and tied up. The men who mount guard must be [have their hair] powdered with flour."[1] Camp cleanliness was also a high priority. Parade grounds were cleared of litter each week before church services in the academy building. This routine continued even after discipline had deteriorated toward the end of the encampment.

Drunk and Disorderly

After 8:00 p.m., a complete transformation took place in the camp. The clean and orderly place became a wild, undisciplined mob scene. Drunkenness and gambling continued unabated through April. Court martial charges typically involved drunk and disorderly conduct with disrespect and disobedience of noncommissioned officers. But it was a rare occasion where the punishment was actually carried out. Lamb's Orderly Book 3 for April 20, 1779 shows that Timothy Donavan was convicted for "selling liquor in camp, absence without leave, riotous and disorderly behavior, and insolence and contemptuous behavior when ordered under guard" Donavan was given thirty-nine lashes and was reduced in rank.

Little was done to prevent the rampant gambling that continued until the encampment ended. Playing cards for money was widespread. Cockfights took place frequently. Knox again admonished the artillerymen and their officers with a vague threat, but this was never applied. A month after the troops arrived, games of chance became so common that Knox issued a specific order to prohibit them:"The pernicious practice of gaming, particularly at cards, prevails among the soldiery...if any soldier, artificer or waggoner shall hereafter be detected of this crime they shall be immediately punished with the utmost severity."[2]

Discipline continued to decline at Pluckemin during the last months of the encampment in the spring of 1779. John More, a matross, and Benjamin Hunt, a gunner, were charged with being absent without leave, disobeying orders and abusing civilians. No evidence can be found that they were punished for these acts. Many men were breaking out of the campground during evening hours to visit the taverns in the village of Pluckemin or to procure liquor from civilian merchants near the camp. It eventually became necessary to increase the number of guards to restrain offenders.

There is much evidence of this decline in discipline in the orderly books. A directive was also issued to stop straggling. Straggling involved either not attending or appearing late for drills and daily musters or work parties. Many other offences were recorded at Pluckemin during the last months of 1779. These including sleeping on guard duty, disobedience, insolence, destruction of civilian property, including burning of fence rails, and striking non-commissioned officers.[3]

Clothing

The artillery regiments at Pluckemin depended on their home states to supply clothing. In all the major winter encampments some state units fared better than others. At New Windsor in 1782 troops from New York, New Jersey, and New Hampshire were so ragged they were billeted in a remote area of the campground where they could not be seen by visitors. But the Virginia companies at Pluckemin were especially well supported. When they first moved into barracks in October they received 550 pairs of stockings, and in the months that followed abundant supplies of shirts, hats, buttons and cloth arrived. Lamb's New York companies fared less well. Their long-awaited shirts did not arrive until March, and when they finally arrived the garments were infested with lice could not be used.[4]

Since camp followers depended on their soldiers for sustenance Molly must have had a difficult winter at the artillery camp. Evidence of this is Colonel Proctor's plea to the Pennsylvania Council for clothing for his men in December: "Ought not the officers and men be furnished with the stipend of clothing that was granted to the rest of her troops as were furnish with the monies to fill the battalion and its complement of men...the honorable house [must acknowledge] the distess of the officers for clothing and other present necessities."[5]

The problem for Proctor's unit was that it was attached for the winter to the Continental Army and therefore Pennsylvania legislature denied responsibility for providing supplies. Finally, in March, after all officers threatened to resign, a shipment of clothing arrived.[6]

The weather turned frigid at the end of the year once all the regiments had arrived in the camp. Knox commented that December 24 was colder than it ever was in Lapland.[7] Many men were without shoes. Good news came on the day after Christmas. A hundred pairs of shoes arrived from New York and were distributed to the men who had the greatest need. A Regimental Orderly Book for December 31 warns that the men that had been given new shoes would no longer have the excuse of avoiding work details.

Food - Feast and Famine

Food for the camp came from the commissary at Raritan, the same source that supplied the main army at Middlebrook. This facility was managed by Assistant Commissary General Royal Flint. As much as 35% of the food supply came from New Jersey.[8] Flour was always in short supply. As early as November 1778 Knox wrote to General McDougall: "We depend on Pennsylvania and southward for flour. The cartage is so immense and expensive that necessity compels us to be as near to the supply as possible."[9]

The flour supply was further limited by a British naval blockade that prevented shipping from entering Maryland ports. Interstate disputes also created additional problems. Four shipments of rice destined for Middlebrook were impounded in South Carolina in retaliation for Congress blocking grain shipments to the southern army.

Meat for Pluckemin came from the central slaughterhouse in Bound Brook. Cattle were driven here from the surrounding countryside, or from eastern Pennsylvania. Curiously, a surplus of meat was a problem at one time. Forage was scarce so that when cattle arrived at Bound Brook they could not be fed. The only option was to slaughter the cows as soon as they arrived. The fresh meat then had to be immediately salted for preservation. This led to a shortage of barrels. As a result, spoiled beef was being delivered to the artillerymen in April. After a large vat was built in April for salting purposes, there is no further mention of this issue in any military documents.

The Role of Pluckemin as a Military Hospital

Sick artillerymen at Pluckemin were treated at the main hospitals at New Brunswick and Bound Brook. However, many were cared for by regimental surgeons in the camp itself before being transferred to the hospitals. A number of huts, isolated from the barracks, were set aside on the north line of the existing structures. Unfortunately, there is no information in military records relating to the number of sick and wounded at Pluckemin available. Regimental surgeon Samuel Adams of the 3rd Continental Artillery was assigned to the camp and did not keep records.

In 1779 the number of sick and wounded at the Middlebrook Encampment overwhelmed the hospitals, and soon they were being housed in barns and public buildings. When Knox's Brigade began leaving the camp in June 1779, General Washington saw an opportunity to end this deplorable situation. He ordered all the disabled who were not in hospitals to be moved to empty barracks at Pluckemin.[10]

The Pluckemin Hospital began to grow rapidly. It accommodated ninety-eight patients by October 1779, when the army began to move into Jockey Hollow for the 1779-1780 winter encampment. From February to May 1780 the Pluckemin Hospital was the largest of eight principle hospitals in the mid-Atlantic area and one of the Continental Army's major hospitals in the northeast. Most of the patients were the wounded returning from the Sullivan-Clinton Campaign against the Iroquois Nations. Many were listed as having chronic infirmities or venereal disease.

Curiously, the hospital had a high desertion rate that far exceeded the level for that of troops on active duty. Apparently, it was easier to abscond from a hospital than a regimental camp. Patients at the hospital were also disorderly. General Washington wrote of the necessity "to prevent and quell all disorders and riots at the hospital."¹¹ Extra guards were posted to maintain control of inmates. The hospital at Pluckemin operated until July 1780, which extended the occupation of the camp for a year after Knox's departure.

Officers Housing Problems

When the first regiments arrived at Pluckemin the officers pitched tents among the common soldiers, but they soon began finding more comfortable quarters living in the homes of area residents. When the last units arrived, they found that all civilian housing had been occupied. The only quarters available were distant from the camp, and several were so faraway that officers did not come to camp or attend drill and parades.¹² A few officers remained in tents, where their living conditions were much more comfortable than for the enlisted men. Artillery officers lived in large marque tents that were spacious and warm.

Preparing for the Next Campaign

Plans for rejoining the main army and its campaigns put an end to the pattern of life at Pluckemin. By March 1779 the daily routine at the artillery camp began to abruptly change. The focus was shifted to preparation for the next campaign and making arrangements for breaking up and abandoning the camp. Knox's Brigade had to be fully equipped to return to the field and transformed into a mobile fighting force once again.

In April General Greene sent orders to Pluckemin that all officers were to be ready to leave their quarters on short notice. He also requested them to toughen up by practicing marching so they would be able to trudge along with their men rather than arrogantly ride alongside them or sit in wagons. The greatest challenge at that time was a shortage of horses to haul the heavy cannons. Knox hurried away from Pluckemin, dragging the heavy guns north over rough roads through Morristown and Pompton to reach West Point, over sixty miles away.

After leaving Pluckemin Lucy Knox moved to Mount Vernon to become a companion for Martha Washington. She envied the fine house and life at the Virginia plantation but yearned for a home of her own, and in letters to her husband she often fondly reminisced about the time they spent at the Vanderveer House.

At West Point Knox gathered the largest formation of artillery ever assembled by the Continental Army, and then directed its transport to Yorktown, Virginia for the final decisive battle of the war. In October 1781 General Knox demonstrated his brilliance. The accuracy of his guns devastated the British forces penned up on the Yorktown Peninsula. Eight days after his artillery opened fire, British General Charles Cornwallis surrendered. Knox's reward was the promotion to major general at age thirty-one, the youngest of that rank in the army. Following the adoption of theUnited States Constitution in 1789 he became President Washington's Secretary of War.

On January 2, 1795, he left the government and returned to his home inThomaston, Maine to care for his growing family. He spent the rest of his life engaged in cattle farming, ship building, brick making, and real estate speculation. Knox died at his home on October 25, 1806, at the age of fifty-six, three days after swallowing a chicken bone which lodged in his throat and caused a fatal infection. He was buried on his estate in Thomaston with full military honors.

Chapter Twenty-Seven

Pluckemin Rediscovered Then Lost Again

Archeology at the Camp

THE MOST OUTSTANDING accomplishment of Knox's Artillery Brigade at Pluckemin was the construction of a large complex of military structures in less than six months. This was the only time in the eight years of the war that a task of such magnitude was performed by an army unit in the field.

What motivated General Knox to build a temporary winter camp on such a grand scale? In this writer's opinion, it was his obsession with promoting the role and the image of artillery in the Continental Army. An impressive cantonment would enhance the importance of the artillery force and gain respect for heavy weaponry as an essential component of the army. At Pluckemin he had the opportunity to prove to the rest of the army and to Congress that cannoneers were not only essential on the battlefield, but that they could rapidly complete major construction projects that required special skills.

The layout of the Pluckemin camp was influenced by Knox's studies of European military science. It's design resembled the Royal Artillery Academy at Woolrich, England, more so than any the other large American cantonments—Valley Forge, New Windsor or Jockey Hollow. The centerpiece of the facility was the academy building from which five buildings extended between 200 and 450 feet. Many of the structures had glass windows and wood floors. Each company, consisting of about sixty men, shared an apartment of two rooms separated by a fireplace.[1]

Knox also recognized the opportunity to promote the spirit of unit pride and esprit de corps in the Artillery Brigade. After a six-month respite away from combat conditions, teamwork on the iconic construction project would provide a unique occasion to instill professionalism and a sense of elitism among his troops.

In modern day warfare, this attribute, known as unit cohesion, is regarded as critical to military success. It was one of the main reasons why the Continental Army did not disintegrate in the latter years of the war. Pluckemin was also an opportunity for Knox to demonstrate the value of familiarizing officers with martial skills. He achieved these objectives. When his brigade left the winter camp it was well trained and equipped. The effective use of artillery proved to be the major force in achieving the American victory at Yorktown and most other prior engagements during the final three years of the war.

The Artillery Brigade Departs

The Pluckemin Encampment began to break up at the end of May 1779 when the artillery force began preparing for new campaigns. As infantry regiments left Middlebrook and marched north, artillery companies were detached to join them. On June 3, 1779 General Knox and his staff left Pluckemin and headed north thirty miles to Pompton, another extensive New Jersey cantonment in Passaic County.

A few activities continued at the nearly deserted Pluckemin location. The facility was placed under the command of Captain Samuel Hodgdon, along with a detachment of guards. They remained there to watch over a large store of ammunition that was left behind in magazines at the campground. The artificer shop also remained open, where wagons were repaired and armorers continued to mend broken muskets.

Captain Hodgdon was assigned to gather the many wagons and teams needed to move the large quantities of supplies left at Middlebrook closer to the main army. These materials included thousands of muskets and bayonets, as well as ammunition. It proved to be a difficult task. On June 12 Hodgdon wrote a letter to General Knox about his problems. The irate officer lamented, "My command here, if it may be called one, has been as difficult as any I have ever experienced. No commissary or provisions left and at least 200 mouths to feed and had I not immediately exerted myself, have reason or believe. I should have been made a prey to their voracious jaws."[2]

General Henry Knox. *(Charles Willson Peale, at Middlebrook, 1778)*

With few guards remaining, the local inhabitants began breaking into the abandoned buildings to steal supplies. Hodgdon requested assistance from Morristown, and then armed his wagoneers to fight off the pilferers. This illicit activity ended when empty buildings became filled with the sick and wounded, as Pluckemin became one of the three army hospitals in New Jersey during the remaining six months of 1779.

After the army completely abandoned the camp, buildings were scavenged and carted away by local civilians for building materials and firewood. Soon there was little evidence of the proud artillery camp remaining, and the once active village of Pluckemin reverted back into a tiny obscure country hamlet. The location of the camp was eventually forgotten. It remained lost for the next 137 years. Even local residents and landowners were uncertain as to its correct location.

The Artillery Camp Rediscovered

Around the turn of the 19th Century, Grant Baker Schley, an affluent landowner in the area, purchased land on the high ground above Pluckemin. He had a vision of opening a resort on the site with a

Max Schrabisch, pioneering archeologist at Pluckemin.
(Courtesy of Trailside Museum, Bear Mountain, NY)

Revolutionary War theme. In 1916 he retained Max Schrabisch, an archeologist and author, to search for the remains of the Knox artillery park. Schrabisch, born in Germany in 1869, was inspired by the findings of archeologists in France and Germany who were discovering the remains and artifacts of the Neanderthal civilization. In 1900 he traveled to America hoping to make similar finds in the northeast United States.

In the years that followed Schrabisch located and excavated hundreds of prehistoric sites in New Jersey and Pennsylvania. The excavation methods at that time were primitive, and much of the evidence he uncovered was not methodically analysed. However, he zealously published the results of his efforts in various books and publications. By the time of his death in 1949 he had written 150 books, newspaper columns, and journals. Even today his works are used as reference guides by archeological researchers.[3]

When Schrabisch arrived at Pluckemin, evidence of the former campsite's possible location could only be found in references in old property deeds, wills, and other legal documents. The land was first owned by John Johnson, who immigrated to America in 1685 and acquired the land in 1701. After his death in 1732 the land was sold to Jacob Eoff. This purchase of 500 acres included the crossroads at Pluckemin, then called Bedminstertown. Jacobus Vanderveer purchased part of the land from Eoff in 1743.

Eoff built a tavern on his property at the crossroads in 1750. Today, the site is at the intersection of Washington Valley Road and Route 202. There was confusion as to whether Eoff or Vanderveer owned the land where the Artillery camp was later discovered. When boundaries of the camp were plotted by archeologists in later years, it was determined that it actually spanned both properties.

In 1917 Max Schrabisch began exploring the heavily wooded hillside above the village of Pluckemin. He discovered twenty mounds that appeared to be the remains of soldiers' huts. The huts had fireplaces, as indicated by ashes, charcoal, and rocks discolored by fire. Animal bones, nails and pottery strewn about the area provided additional evidence of the shelters. All mounds contained bottle glass and pottery fragments.

Twelve of the hut sites were excavated. Most of the artifacts that were discovered related to the daily lives of the artillery soldiers. Foods consumed included oyster shells, turkey and bear bones, proof that the troops hunted to supplement their daily rations. Military relics that were discovered included lead shot and buttons marked with regimental numbers and the initials "U.S.A." A sword belt tip showing a canon and a flagstaff was also among the objects unearthed.

Schrabisch also located what appeared to be the site of a large blacksmith shop, where he found ox and horse shoes, nails, hooks, and sheet iron. It is likely that this facility was the artificers and armorers shop where wagons and guns were repaired. After discovering the location of the campground, he hastily attempted to restore the remnants of the structures and then to stabilize their remains. His ambitious archeological project took place over a course of eleven weeks.

Unfortunately, Schrabisch did not keep field notes or leave maps to indicate where the artifacts were discovered. His inept restoration work impeded future efforts by increasing the difficulty to distinguish his repairs from original ruins. In the years that followed the extensive collection of artifacts he discovered was lost.

The best evidence of the Schrabisch project is found in a series of 1917 newspaper columns in the local *Bernardsville News*.[4] Schrabisch's work stopped soon after his sponsor Grant Schley died in 1917. The site was soon forgotten, and the clearings and excavations left by the work of Schrabisch were once again covered by forest duff. Only a few local residents remembered the existence of a Revolutionary War site, describing it as "somewhere up on the hill."

After Schrabisch completed his project in 1918, the artillery campsite at Pluckemin was forgotten. It languished in obscurity for the next forty years. In the 1960s, Somerville resident Clifford Sekel was working at the National Archives on a master's thesis for Wagner College. He noticed frequent referrals to Pluckemin in documents from the Revolutionary War era.

He correctly surmised that the Pluckemin Artillery camp, almost unnoticed in history books, had been overshadowed by the previous winter of suffering at Valley Forge and the harsh season at Jockey Hollow the following year. Yet Pluckemin had more immediate and lasting impact on the course of the war than either of these places. Other than military documents, the inquiring student was unable to find any further evidence or detailed information about the size of the camp or its role in the war. He was amazed that no New Jersey historian or resident of the Pluckemin area remembered the precise location of the historically significant site.

A Curious Scholar

The inquisitive graduate student became obsessed with learning more. In the summer of 1966, he began hacking his way up and down the mountain with a machete looking for evidence of structures that could verify the plethora of clues found in the military records. Sekel was apprehensive, since the land was marked for residential development and logging which could potentially destroy the site. He combed the mountainside for twenty weeks over two summer vacations in the 1960s without finding any evidence of military presence.

In 1967, Anne O'Brien, a reporter for *The Bernardsville News*, discovered the 1918 articles written by Max Schrabisch in the newspaper's archives. The accounts identified a key feature, an uncovered garbage dump that contained thousands of clam and oyster shells. Sekel had noticed the pile before but assumed it was of Lenape Indian origin. When he began clearing the area surrounding the heap in late 1967 he soon found the outlines of buildings. He knew that he had rediscovered the site of the artillery camp.

It took Sekel another ten years to arouse enough interest to secure the site for the non-profit Pluckemin Archaeological Project. During that time he was supported by the Hills Development Corporation, Bedminster Township, and scores of small businesses, foundations, and individuals.

Sekel was joined in field work by Rutgers archeologist John L. Seidel in the late 1970s. Seidel's professionalism and scrupulous field work over the next seven years provided the bulk of what is now known about the construction of the Pluckemin site. Seidel conducted the first systematic survey of the site in 1979. Sekel and Seidel were joined by Bruce W. Stewart, a historical geographer and park historian at the Morristown National Historic Park. Seidel's work is described in a dissertation written for his doctoral degree at the University of Pennsylvania in 1987.[5]

Professional Assistance and Financial Backing

The team started the archeological dig in 1979. After clearing the foliage in the area, a series of grid points was set across the site. A team directed by Sekel and Seidel assisted by students from Drew University, Somerset Community College and Rutgers University started the archeological dig in 1979 which continued through 1986. A detailed map of landscape features with the locations of discovered artifacts was prepared.

The archeologists were guided by the 1779 drawing of the Pluckemin Cantonment by Continental Army Captain John Lillie. Lines of stone and old walls identified the line of barracks and the probable location of the academy. Several advanced technologies were used to document the site. Low level aerial photography was used to record patterns of artifacts. Ground penetrating radar and magnetometry were employed to detect variations in soil magnetism that would indicate archeological features.

The thousands of artifacts found told a fascinating story. A rusty bayonet, and a belt buckle decorated with a cannon and thirteen-star flag were unearthed. The belt tip is the earliest known artifact to clearly show the American flag. It was engraved by a Philadelphia silversmith who was called to the camp in early 1779. It is likely that he etched the flag based on what the army was actually displaying in camp at the time. It proves that the Continental Army was using the "Stars & Stripes" flag in that year. This is the earliest provable American flag design, with thirteen stars laid out in a pattern of five rows.

The 190,000 artifacts that were eventually uncovered provide a rare insight into the daily lives of Revolutionary War soldiers. Monmouth University archaeologist Richard Veit later joined Seidel to analyze the extraordinary collection unearthed at the location. He described the Pluckemin site as one of the great overlooked stories of the American Revolution, with evidence that provides an unparalleled glimpse of the lives of Revolutionary War soldiers.

Artifacts revealed that a forge was located in the southeast part of the site. Objects associated with the preparation of food were clustered in the northern area. Officer's quarters could be identified by their trash deposit, which contained a higher concentration of expensive porcelain and imported English tea wares than those found in deposits near soldier's quarters.

The large number of artifacts unearthed during the project may be the greatest collection from a single Revolutionary War site in the country. While the archeologists worked on the excavations, the developer, Hills Corporation, began construction of condominiums in areas where the evidence of the campsite would not be disturbed. The archeological project continued until 1989 when it was halted due to lack of financing.

The Jacobus Vanderveer House, following its restoration by The Friends of the Jacobus Vanderveer House. *(wikipedia.com)*

The extensive artifactcollection was placed in storage after the completion of excavations. An analysis of its contents was nevercompleted. In 2007, John Seidel, the director of the project in the 1980s, joined with the Friends of the Jacobus Vanderveer House, Hunter Research, a cultural resource consulting firm, and Monmouth University to revive the project.

Their objective was to re-examine the entire artifact collection to create a database and to prepare publications that would serve to gain broader public recognition for the Pluckemin site. Plans were made for theVanderveer House to be the repository for the artifact collections and records as well as to serve as an interpretive center for the cantonment. In 2012, Seidel issued his first report on the history and archeology at the Pluckemin ContinentalArtillery camp. Today, researchers are still cataloging about a million artifacts that were recovered.

The decade of excavation produced one of the largest and best recorded collections of Revolutionary War artifacts in the country. Some portions of the site were subsequently built on by the Hills Development Corporation. These areas were archaeologically investigated to ensure that they were devoid of artifacts. The bulk of the site is currently owned and preserved by Bedminster Township.

In 2016 the Friends of the Jacobus Vanderveer House & Museum made plans to move a decaying, but architecturally significant, early nineteenth century Dutch barn from Branchburg, New Jersey to the grounds of the house. The 33' x 51' barn on Old York Road, when reassembled, will provide a secure repository for the archaeological artifacts that were excavated during the Pluckemin Archaeology Project. These artifacts and the accompanying documentation are currently being housed in a warehouse facility in Central New Jersey until they can be permanently stored in a climate-controlled facility and be available for exhibition and interpretation.

An Amazing Discovery - The Lillie Drawing

In 1979 Sekel and Seidel discovered an amazing document in the archives of Morristown National Historical Park. Files there made reference to an 18th century drawing of the Pluckemin Artillery Camp. But the drawing was missing from the collection. Curiously, it was reported to be in the possession of a park employee who claimed to have obtained it at an auction. The historians were astonished when they were able to view it. The remarkable drawing depicted the same parallelogram of buildings that had been described in the *Pennsylvania Packet*. The remarkable sketch was signed I. Lillie and titled *A South-West Perspective View of the Artillery Barracks, Pluckemin, N. Jersey*.

The plan was drawn in pen and ink and embellished in watercolor. John Seidel described the layout of the camp as it is depicted in the drawing:

> The rear of the parallelogram (the east side) is comprised of two long buildings that extend lengthwise from north to south, and east to west. Uphill to the rear of the northern building, is a large square structure with two chimneys. The north and south sides

Pluckemin Rediscovered 311

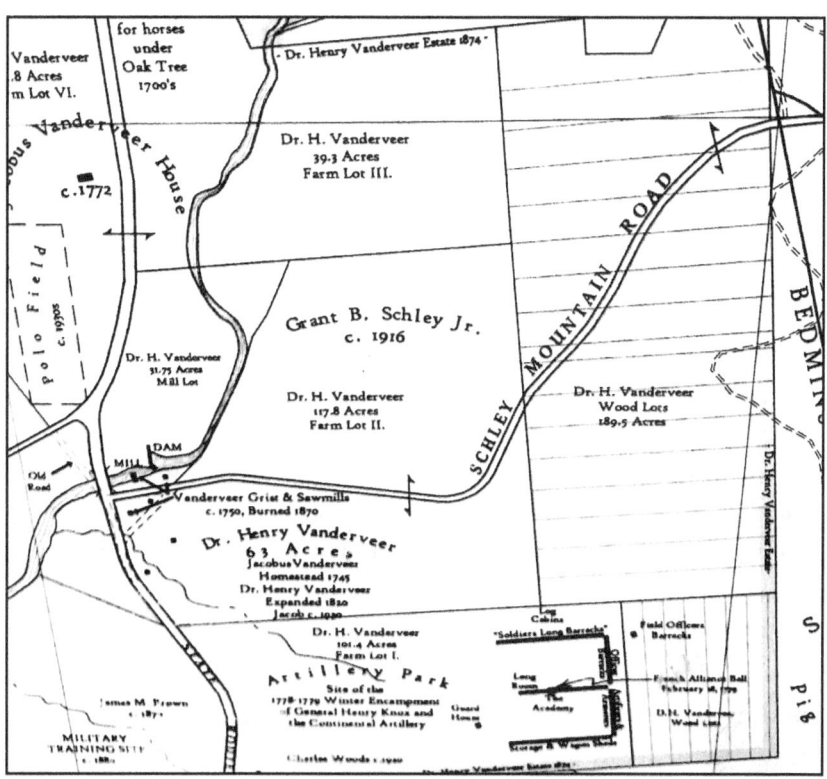

With over 2,000 hours of painstaking research the landscaping firm of John Charles Smith & Associates of Far Hills has prepared this historic look back at the history of Bedminster Township. The rendering highlights historic property lines, roads, buildings, and the lost Pluckemin Artillery Park. While the map states it covers ownership from 1893-1939, there are obvious land markings that go back to the early 18th century. The map, including the Vanderveer property highlighting the lost Pluckemin Artillery Cantonment area, is on display at the Jacobus Vanderveer House in Bedminster, New Jersey. *(Courtesy of The Friends of the Vanderveer House)*

of the parallelogram each consist of a long east-west building, both facing the rear of the parallelogram. Down the center is a fifth long structure punctuated with a large cupula type building towards the center. The building corresponds to the description of the academy building. To the north of the parallelogram is an apparently random scatter of huts and a somewhat larger cabin is situated south west of the probable academy line.

Seidel claimed that the there is no doubt that the image is authentic. On the lower right hand corner is inscribed I. Lillie facit. Captain John (Ionas) Lillie commanded the 12th Company of the Third Continental Artillery, and is known to have been at the camp. The fact that the details of the drawing correspond with the subsurface archeology uncovered in the excavations help verify its authenticity. Seidel also observed that the parallelogram pattern at Pluckemin is very similar to that recommended by the military plans of the time for artillery camps. In general, artillery tent camps were formed with enlisted men along the sides, guns in front, and officers in the rear. Knox followed this plan, but with permanent structures.

The Artillery Brigade force of 700 to 800 men, commanded by Brigadier General Henry Knox, functioned independently, and established the cantonment in Pluckemin. The camp had a profound influence on the conduct of the war, a conflict that continued for almost three more years. The effective use of artillery with American gunnery expertise was a major contributor to victory. The academy. the officer training facility that operated twenty-four years before the founding of the United States Military Academy at West Point, distinguishes the Middlebrook from the other major cantonments of the Revolutionary War.

Chapter Twenty-Eight

Conclusions and Epilogue
Middlebrook and Pluckemin Remembered Today and into the Future

AT MIDDLEBROOK THE Continental Army reached maturity. The gathering of disparate colonial militias under General Washington's leadership developed into a cohesive and organized fighting force during the winter of 1778-1779. The image of a raw army of young farm laborers, untested and unmotivated, with short enlistments and with allegiance only to their states, vanished at Middlebrook. The men who had survived a crucible of bitter defeats over the past three years had become resilient professional soldiers with the realization that it was possible to win the war.

Washington and his generals built a unified professional military organization that came together at Middlebrook and ultimately enabled the Continental Army to triumph. Throughout the encampment, Washington inspired the soldiers through his own dedication and effort to influence Congress to reform the supply system and end the crippling shortages.

Von Steuben continued to teach the soldiers military skills and to fight as a unified army. These reforms, in supply systems and fighting tactics, when combined with improvements in military hygiene and army organization, became the foundation of the modern United States Army. Private Joseph Plumb Martin expressed it best when he proudly described the rejuvenated army, writing "We had engaged in the defense of our wounded country and...we were determined to persevere."[1]

Less than six months before the Middlebrook encampment,

The author at the Martinsville Redoubt. *(Author's collection)*

the Patriots had driven the British regulars, acknowledged to be the best army in the world, from the field at Monmouth Courthouse. American commanders had become stronger and more confident.

Seasoned veterans who had served for the previous three years and had enlisted for the duration of the war were proud their professional fighting skills. The American Army at Middlebrook eventually resembled the best European forces and could be ranked with their British adversaries or the Prussian forces of the brilliant campaigner Frederick the Great.

The successful encampment at Middlebrook that winter made possible this resurrection and maturing of the Patriot forces. Close to the enemy nerve center in New York, Middlebrook was a naturally defensible bastion where troops could train and recoup from the year's battles when winter weather, impassible roads, and scant supplies stopped the fighting. The location of the camp deterred

Conclusion and Epilogue 315

British commander Henry Clinton, with half of his army in New York, from attacking American supply sources in New Jersey.

Lessons learned from the disaster at Valley Forge in 1777-1778 led to improved hutting and sanitation during the second Middlebrook Encampment in 1778-1779. Fortunately, the weather that winter was mild, although some snow fell in late December and early January. After that the temperature generally remained above freezing. It warmed in the spring weeks earlier than usual, with fruit trees flowering during the first week of April.

During the winter, Quarter Master General Nathanael Greene managed to keep the troops fed and clothed. At times there was hunger in the camp but never at the level of the widespread starvation at Valley Forge the year before, or at Jockey Hollow a year later. The alliance with France created a great optimism that was sensed by even the lowest private. In February new uniforms supplied by the allies were distributed to most of the men. Incredibly, a surplus of donated clothing required storage facilities to be built. A grand review of the entire Continental Army was held in honor of a visit by the French ambassador in March 1779.

The demographics of the army had almost completely changed since 1775, the first year of the war. Few land-owning yeoman farmers with families remained in the regular army. At Middlebrook most of the soldiers were young, poor, unmarried farm laborers. What originally was an army of mainly New Englanders now encompassed men from all parts of the new nation. Half of the soldiers were from the mid-Atlantic states. Pennsylvania had a "German Battalion," and many other regiments were composed of forty percent Scots-Irish Presbyterian immigrants from Ulster or their descendants.

While at Middebrook, General Washington planned and gathered supplies for the campaign against the Iroquois Nations in New York State. In late May the troops began to be ordered away from New Jersey to new posts. The last regiments left during the first week of June, The operation against the Native Americans and their Loyalist allies, led by General John Sullivan, included the New Jersey

Winter, Middlebrook 1779 *(Courtesy U.S. National Postal Museum)*

Brigade. The huts at Middlebrook were abandoned for the next five months until their destruction became the objective of a British raid in October 1779.

At that time Colonel John Simcoe, commander of the Queens Rangers, an elite unit of the British special forces of mounted dragoon troops, led a raid from New Brunswick into the Raritan Valley. After setting fire to barges and pontoon boats at Van Veghten Bridge, which today crosses over the Raritan River into Manville, he burned the Dutch Reformed church and courthouse at Millstone. When the raiders reached the huts at the Middlebrook camp sites they found that the timber was too damp to ignite. Simcoe attempted to withdraw to avoid being trapped but was ambushed and captured by militia forces.

In August 1781, two years after Simcoe's raid, Middlebrook again came to life when the allied army, a force of more than 16,000 American and French soldiers, passed through the camp on the way to Yorktown. Later that year, the allied forces again visited the camp when they returned north.

That was the end of any military activity for the Middlebrook Encampment. During the years after the war the huts were torn down by local residents and used as building material or firewood. Evidence of the structures remained as late as 1837 when Reverend Abraham Messler of the First Church of Raritan hiked up Vossellor Avenue to visit the site of Wayne's campground on the first ridge. He reported

Conclusion and Epilogue   317

that remnants of huts still remained. A large earthwork redoubt that guarded the pass at Chimney Rock can still be visited in Martinsville.

A guide prepared in 1939 by the Works Progress Administration (WPA) reported that on Route 22, along the northern fringe of Bound Brook, there were remains of "breastworks thrown up by Washington's men." These were described as low hummocks covered by foliage on both sides of the road. The highway was widened in recent years and no evidence of these mounds remains.[2]

Today, much of the ground where the Middlebrook encampment once stood lies under Routes 22 and Interstate 287. A flagpole now marks the center of a twenty-acre park on Middlebrook Road off Vossellor Avenue, north of Route 22, in Bridgewater Township. The Washington Campground Association dedicated the park as a historic site in 1889 and it was entered in the National Register of Historic Places on July 3, 1975.

The campground was spread over more than five miles. Most of it has been built over by residential and commercial development and highways. A detailed, systematic, and thorough assessment of the archaeological potential of the land has never been undertaken. Large tracts of unexplored land lie within Somerset County's Washington Valley Park. This publicly owned land includes much of the acreage used in the first encampment of 1777. It has never been explored or developed. Archeologists agree that much substantive archaeological evidence still remains relating to the Middlebrook encampments in Bridgewater Township and its environs.

A project to archaeologically explore sensitive locations would increase public awareness of the historical importance of the two Middlebrook encampments. It could also encourage state and county open space acquisition programs to assign a higher priority to the purchase of sites that hold a high potential for yielding significant archaeological evidence. Historic preservation might be incorporated into local land development decisions. The Washington Valley, Bound Brook vicinity, and vantage points along the First Watchung ridge could be promoted as tourism destinations.

Extensive archeological studies have been done only at two key locations. A historical and archeological study of Washington Valley Park east of Middle Brook and Wayne's brigade encampment in 1777 was completed in 2003 by historical resource consultants at Hunter Research. Archeological digs were made at the Artillery Park at Pluckemin over several years, resulting in thousands of artifacts recovered.

Middlebrook passes almost entirely unacknowledged in contemporary media or historical literature. Buried, obliterated, or concealed by suburban sprawl, the encampment area is again fading from memory as a national Revolutionary War heritage site. Public awareness should be increased, and a greater effort made to present the history and archaeology of the site to the public. It should be ranked alongside the nationally acclaimed Revolutionary War encampments as Valley Forge, Jockey Hollow and New Windsor.

What Henry Knox and his men accomplished at the artillery camp at Pluckemin was hugely impressive. He firmly established the necessity of organized training for officers and founded the nation's first military academy. Barracks constructed there were large, and substantially built. The archaeology study of the site has enhanced the appreciation and understanding of these accomplishments. It has provided an authentic account of the everyday life of officers, enlisted men and civilian workers.

The grounds of the artillery campsite at Pluckemin are not currently accessible to the public. There are no buildings or trails in this heavily wooded area and most of the site has been intensively developed with residential housing and surrounded by the Hills condominium development. Other sites related to the encampment, in Bridgewater and surrounding towns, are accessible to the public. These include Washington Rock State Park, the observation site on top of the first Watchung Mountain ridge in Green Brook, and nearby Washington Valley Park in Bridgewater.

Historians have also failed to recognize the continuous importance of Middlebrook throughout the tumultuous Middle

Conclusion and Epilogue

Atlantic campaigns that took place between 1777 and 1779. Throughout this period the New Jersey Militia based at the camp played a vital role by immobilizing the Crown forces and signaling their movements.

Thousands of commuters drive past historic Middlebrook sites each day but have no appreciation of the many critical events of the Revolutionary War that occurred there. The encampment is one of the great overlooked stories of American history. Unfortunately, the area has only a few historical markers denoting New Jersey's history. Further development is constantly encroaching on the remaining vacant ground.

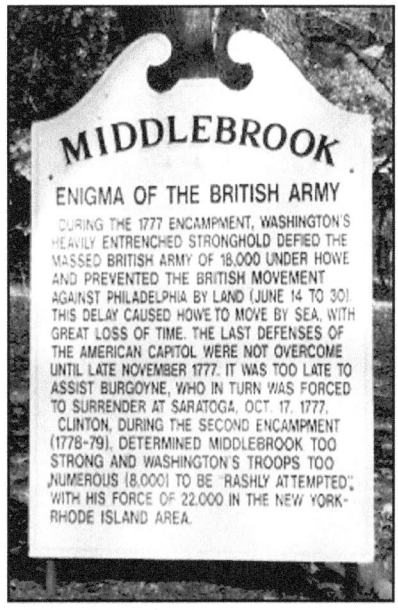

Signage on 20-acre Washington Campground Park, 1761 Middlebrook Road, Bound Brook, NJ. *(wikipedia.com)*

Appendices

Appendix A

Timeline of Middlebrook Encampment and Related Military Activities

1777

Jan. 3	Washington's Army moves from Princeton to Morristown
Jan. 4	British Army moves from Princeton to New Brunswick area
Jan. 21	Forage Wars begin with Battle of Millstone or Van Nest's Mill
March 1	Bound Brook Outpost established
April 13	Battle of Bound Brook
May 19	Continental Army moves from Morristown to Middlebrook

First Middlebrook Encampment
May 19, 1777 to July 4, 1777

May 28	General Washington sets up Headquarters in Martinsville.
June 13	British Army advances from New Brunswick to Millstone.
June 14	Washington's HQ moves to Waynes Brigade campground on first ridge
June 22	British Army retreats to Amboys from New Brunswick. Washington moves army to Samptown.
June 24/25	Stirling deployed to Edison, Scotch Plains, (The Short Hills) British Army reverses course to trap American army at Samptown Americans avoid trap. Army moves back to Middlebrook. Washington moves his lookout to Green Brook rock. Washington at Vermeulen House, Plainfield. Washington at Drake House, Plainfield.
June 26	Battle of the Short Hills. Woodbridge, Edison, and Scotch Plains

Appendices

June 26-28	Stirling retreats from the Short Hills to Middlebrook. British withdraw through Westfield and Rahway.
June 30	British Leave New Jersey for New York.
July 4	American Army leaves Middlebrook for Hudson Highlands.

1778
Second Middlebrook Camp October 1778 to June 1779

Feb. 6	Signing of Franco-American Alliance
Oct. 29	Middlebrook selected for main winter campsite.
Nov. 1-15	Patriot forces begin March from New York to Middlebrook Supplies forwarded from Philadelphia to Middlebrook.
Nov. 23	Hut construction begins.
Nov. 29	Gen. Greene arrives to assign regimental camp sites
Dec. 11	Gen. Washington arrives and sets up HQ at Wallace House.
Dec. 4	American line of march is threatened from New York.
Dec. 7	Artillery arrives at Pluckemin, 675 troops, 60 cannons.
Dec. 9	Maryland and Virginia Regiment arrive.
Dec. 15	Pennsylvania troops arrive.
Dec. 22	Gen. Washington leaves camp to visit Congress in Philadelphia.
Dec. 23-31	Violent snowstorm and freezing weather delay supply delivery.
Dec. 30	Supplies begin arriving by sled from Morristown.

1779

January	Rolls show 8,875 troops at Middlebrook, 4739 fit for active Inflation curtails the Army's ability to purchase supplies.
Jan. 10	Barracks completed at Pluckemin. Knox moves into Vanderveer House
Jan. 18	British prisoners captured at Saratoga request aid.
Jan. 31	All hut construction at camp completed.
Early Feb.	Knox visits Congress at Philadelphia to appeal for funds.

Feb. 5	Gen. Washington returns from Philadelphia
Feb. 9	Mutiny at Pluckemin.
Feb. 13	Supply of flour in camp reduced to four days.
Feb. 18	New Year's celebration of French Alliance at Pluckemin
March	Rolls show 7029 troops at Middlebrook.
March 19	Officers at Pluckemin threaten to resign.
March 23	Signal Beacon erected high ground behind barracks at Pluckemin.
April	Rolls increase to 9804 troops with arrival of new recruits. General Washington reports, "Treasury at Middlebrook is entirely empty." Corps of Light Infantry organized-2,000 elite troops
April 15	Gen. Greene to Philadelphia, Congress to appeal for funds.
April 20	Benedict Arnold trial for appropriating equip. for his personal use.
April 27	Baron von Steuben appointed Inspector General of the Army.
April 29	Rancid beef causes widespread sickness in camp.
May 1	French Delegation arrives.
May 14	Delaware Indian delegation visits camp.
May 18	Sullivan-Clinton Indian Campaign force leaves camp.
June 3-5	Gen. Washington leaves Middlebrook with his Life Guard; Artillery leaves Pluckemin.
June 3-5	Remainder of 10,000 troops leave Pompton.

Appendix B

Deployment of Continental Army for Encampments Winter 1778 - 1779 Washington's Papers

Head Quarters, Fredericksburg, November 27, 1778.

Sir: The Army is ordered to Winder Cantonments at the following places Viz.

- Parsons............... ⎫
- Huntingtons......... ⎬ to Danbury
- Poors.................. ⎭
- Pattersons............ Fort Arnold at West Point.
- Nixons................. Continental Village near Peekskill.
- Learneds.............. Fishkill.
- Clarks (No. Carolina) Smiths Clove and West side of Kings Ferry.
- Woodfords........................ ⎫
- Muhlenbergs...................... |
- Scotts............................... |
- Smallwoods....................... ⎬ Middle Brook in Jersey.
- Hall ad. Maryland............... |
- Waynes............................. |
- Irwin ad. Pennsylvania....... ⎭
- Dubois............... ⎫
- Vanschaicks....... ⎬ Regts. at Albany
- Livingstons........ ⎭
- Courtlandt....at Rochester, between the Minisink and Æsopus.

On Tuesday next until you receive orders to the contrary, you will direct the officers to march the men which compose your Light Corps to the different Regiments to which they belong by the nearest routes to the above Stations

Appendix C

American Generals at Middlebrook, 1778 - 1779

Commander–in-Chief
George Washington

Major Generals
Baron Johanne De Kalb.. Maryland Division
Louis Du Portail... Chief of Engineers
Nathanael Greene... Quartermaster General
Arthur St Clair.. Pennsylvania Division
Baron Frederick von Steuben .... Inspector General
William Alexander, Lord Stirling............................... Virginia Division

Brigadier Generals
Henry Knox.. Commander of Artillery
Peter Muhlenberg........................... Commander 2nd Virginia Brigade
Charles Scott .... Commander 3rd Virginia Brigade
William Smallwood........................ Commander 1st Maryland Brigade
Anthony Wayne....................... Commander 1st Pennsylvania Brigade
William Woodford .... Commander 1st Virginia Brigade

Appendix D

Somerset County Residents Who Rendered Services for the Continental Army at Middlebrook

Adam Alberger, butchering
Alexander Brenan, carting tallow
John Burner, tending cattle
James Burnett, carting hay
James Burnsides, building defenses
Henry Cooper, driving cattle
Thomas Curtis, carting salt
Henry D'Camp, attending cattle
Charles Dean, butchering
John Doremus, victualing sick
Abraham Drake, victualing soldiers
James Dunham, driving cattle
Daniel Dunn, carting tallow
John Fisher, slaughtering cattle
Jacob Flint, riding express
Aaron Foreman, unknown
Abraham Fowler, butchering
John Garvins, butchering
Arche Gifford, cooper
D. Hagerman, carting flour
Abner Hailfield, butchering
Amanias Halsea, carting hay
Daley Halsea, carting hay
H. Hankinson, carting salt
John Hopper, carting hay
William Howell, driving cattle
William Johnes, Slaughterhouse
Edward Lammon, carting salt
Jacob Lappen, Drover
Nathan Lion, carting hay
Aaron Ogden, butcher
James Polles, taking in pork
John Post, carting hay
John Rogers, butchering
Sam Ryder, butchering
John Smith, butchering
Moses Stiles, carting hay
Thomas Throgmorton, carting salt
Wes Tucker, carting spirits
Nathan Turner, butchering
ParrotVan Deveer, carting to camp
Cornelius Vorheis, driving cattle
Thomas Washington, butchering
Samuel Woodruff, butchering

Civilians of Somerset County employed as express riders for the army 1778-1779

James Dunn, Joseph Burnwell, Nicholas Christopher, James Davis, Methuslah Davis, William Day, Robert Dunn, Isaac Fitzworth, Noah French, Isaac Hempstead, John Long, M. Lyon, Joseph Marsh, Alex McCallister, Moses Ruften, John Turner. (Source: *Middlebrook- The American Eagle's Nest* by Carl E. Prince, Somerset Press, 1958.)

Appendix E

Middlebrook Encampment-Houses of Senior American Officers Self-Directed Tour

This self-directed tour links together the five existing Revolutionary War era houses that were occupied by General Washington and his generals during the Middlebrook Encampment 1778-1779. The route of the tour follows many of the same roads that existed during that time. Your journey can start from any of the five sites and can travel in either direction.

The Abraham Staats House occupied by Baron von Steuben, is located in South Bound Brook just off Main Street. From Bound Brook cross the Queen's Bridge to South Bound Brook and bear left on Main Street 1/2 mile. Turn left onto Von Steuben Lane to the end.

From the Abraham Staats House proceed north on Main Street to the Queen's Bridge crossing the Raritan River. At the traffic circle, proceed west on Main Street, known as the Old York Road, to the **Van Horne House** occupied by General Benjamin Lincoln. It is located on the right just after passing under Route 287 and is across from the Commerce Bank Ballpark. You can Park behind the house.

From the Van Horne House continue driving west on Main Street to Finderne Avenue Turn left on Finderne. Turn right on Van Veghten Drive at the first stoplight. **The Van Veghten House** is located at the end of the road. It was occupied by General Nathanael Greene.

From the Van Veghten House return to Finderne Avenue and turn left. At the next light, turn left onto Main Street. Proceed through Somerville to the "Y" at Borough Hall. Bear left onto Somerset Street. Just after passing under the railroad underpass, Look for the Wallace

Appendices

House Parking lot on the left. This house was the headquarters of General Washington.

From the Wallace House, continue west on Somerset Street to Route 206. Take Route 206 west to Lamington Road. Make a "U" turn and drive south on Route 206. Bear right for River Road and look for a cemetery on the right. **The Vanderveer House** is just beyond the cemetery. You can park behind the house. It was occupied by General Henry Knox, commander of the Pluckemin Artillery Camp.

Appendix F

Sequence of Ownership of
The Oliver/McCain/Campbell Farmstead

The Oliver/McCain/Campbell Farmstead is on the crest of the first ridge extending from Vossellor Avenue along Miller Lane to Chimney Rock. It covers the area that Wayne's Regiment occupied during the 1777 encampment and the site of Middlebrook Washington Rock.

Tenure	Name	Reference
1688-1706	Thomas Codrington	East Jersey Deed B:410
1706	Philip French	East Jersey Deed D2:1
1706	Elizabeth French, et. al.	(Maas 1975:86)
1722-1750	Joseph Reade	East Jersey Deed D2:7
1750-1777	Thomas Muckleworth [McElrath]	East Jersey Deed I2:3
1777-?	Thomas McElrath II	New Jersey Will 19 348
c. 1779-1798	Jeremiah Oliver	Bridgewater Tax Ratables
1798	Martin Oliver John Leonard, Jr.	Somerset County Will 1084R
1799	James McCain	Somerset County Deed B:459
1820-1852	Margaret Campbell	Somerset County Deed J:850
1852-1853	Elias C. Milliken	Somerset County Deed P2:179
1853-1884	Peter Snyder	Somerset County Deed R2:266
1884-1889	Mary Weber	Somerset County Deed H6:273
1889-1890	George Schwed	Somerset County Deed X6:504
1890-1892	George Snyder	Somerset County Deed Y6:143
1892-1893	William H. Cawley	Somerset County Deed L7:275
1893	Maria L. Vanderveer	Somerset County Deed O7:200
1893-1909	Joel Coddington	Somerset Cty Deed O7:208, 2111
1909-1951	Francis Morris	Somerset County Deed C12:365
1951-1970	Robert K. Haelig, Sr.	Somerset County Deed 766:124
1970-2000	BPR Land Devlopment Co.	Somerset County Deed 1226:30
2000-	County of Somerset	Somerset County Deed 2316:525; 2347:109;

Source-Hunter Research Study 2013, 3-33

Appendix G

The Signal Beacons on the Watchung Hills

To defend from a surprise attack by Clinton's forces out of New York, General Washington ordered General Alexander, Lord Stirling to plan a series of signal beacons be built "on conspicuous hills and Mountains, on the eastern side of the Watchung Mountains, then called the Blue Hills. Stirling's home was in Basking Ridge and knew the area well. A total of twenty three beacons were built. The Watchung Mountains are a multi-ridged mountain chain. The first eight beacons were built on the first ridge; the remaining fifteen were built on the northeastern or second ridge. They stretched for over forty miles from Somerset to Bergen County. The location of some of the beacons was very specific while others were atop a nondescript hill or mountain near a town.

The locations described by Baron de Kalb with numbers and site data added for clarity:

"Signals on which the militia are immediately to assemble"*

1. "A Long fire on the Mountain in the rear of Pluckemin"
2. "one on the mountain near steak [Steel] Gap"[Bridgewater]
3. "one on the mountain near Mordicas or Wayn's Gap"[Bridgewater]
4. near Linelons [Lincoln's] Gap"[King George Road, Greenbrook]
5. "one near Quibble Town Gap" [near Washington's Rock Lookout, Greenbrook]
6. "Hill, road to Baskin-Ridge four miles north of Col. Van Horns" [Vossellor Ave]
7. "on the Hill toward Princetown"
8. "one on the Hill in front of Marten Taverns near short hills"

9 "at the point of the mountain [Beacon Hill] north of Springfield one mile
10 "on the top of the Hill [Hobart Hill in Summit?] one mile south East of Chatham"
11 "at Cooper's Wind Mill on Long Hill at [Col.] Ludlows"
12 "at the point of Kennys Hill at Morristown [Fort Nonsense]"
13 "on Pidgeon Hill four miles northwest of Morristown"
14 "on Schuylers mountain N. W. of Pluckinin, 12 miles"
15 "on the Hill 10 miles west of Do."
16 "on the South Point of Cushatunk [Round Hill?]"
17 "on the Hill [Prospect Hill] N. W. of Fleming towns"
18 "on the N. W. point of the Southern Hill [Goat Hill?] "
19 "on the Hight of Amwell looking southward"
20 "one on Princetown looking southward"
21 "on the Carter hill in Monmouth"
22 "on Middleton hill"
23 "on mount Pleasant"

Source: Frédéric, Baron de Kalb to Washington, March 14, 1780, The Founding Era Collection, Founders Early Access; in the Peter Force *Transcripts of Washington Papers (Continental Army Returns,* No. 36, vol. 3, 119) "No. 2. Signals on which the Militia are immediately to assemble" which locates beacons number 1 through 23.

Appendix H

Historic Sites of Somerset County New Jersey

Northern

1. The Boudinot-Southard-Ross Estate, Basking Ridge
2. The Codington Farmstead, Warren
3. The Jacobus Vanderveer House, Bedminster
4. Dr. John Vermeule House, Green Brook
5. The Kennedy-Martin-Stelle House, Basking Ridge
6. The Kirch-Ford-Terrill House, Warren
7. Mount Bethel Baptist Meeting House, Warren
8. Texier House Museum, Watchung
9. The USGA Museum, Liberty Corner
10. The Vermeule Mansion, North Plainfield
11. Washington Rock, Green Brook

Central

1. The Abraham Staats House, South Bound Brook
2. The Andrew Ten Eyck House, Branchburg
3. The Brook Theater/Somerset Cultural Arts Center, Bound Brook
4. The Daniel Robert House/Somerville Borough Hall, Somerville
5. The General John Frelinghuysen Home/Raritan Public Library, Raritan
6. The Old Presbyterian Graveyard, Bound Brook
7. The Presbyterian Church of Bound Brook, Bound Brook
8. Relief Hose No. 2, Engine House, Raritan

9. The Shrine of the Blessed Sacrament, Raritan
10. The Somerville Exempt Fire Museum, Somerville
11. The South Branch Schoolhouse, Branchburg
12. The Van Veghten House, Bridgewater

Southern

1. The Griggstown Bridgetender's House, Franklin Township
2. The Griggstown Schoolhouse, Franklin Township Harlingen Reformed Church, Belle Mead
3. Historic 1860 School House/Millstone Borough Hall, Millstone
4. The Old Millstone Forge, Hillsborough
5. Stoutsburg Sourland African American Museum, Skillman
6. Van Liew-Suydam House, Somerset

Appendix J

*Somerset County Historical
Associations and Commissions*

Somerset Co. Historic Society, Van Veghten House, 9 Van Veghten Dr. Bridgewater, N.J. (908) 218 1281

Washington Campground Association, Meets on Washington's Birthday (Feb 22) and July 4 at commemorative park, 1761 Middlebrook Rd., Bound Brook, N.J.

Heritage Trail Association Van Horne House, 941 E. Main St., Bridgewater, N.J.

Somerset County Cultural and Heritage Commission County Administration Bldg, Somerville (908) 231 7106   E-mail CulturalHeritage @co.somerset.nj.us

The Historic Society of the Somerset Hills PO Box 136, Basking Ridge N.J. 07920 (908) 221 1770

New Jersey Historical Society (Jerseyhistory.org) 52 Park Place. Newark, N.J. (973) 596 8500

League of Historical Societies of New Jersey Linda Barth (908) 722 7428 Barth 123@aol.com

New Jersey Historical Commission newjerseyhistory.org

Endnotes

Chapter 1. Strategic Situation [pages 3-7]

1. Richard Ketchum, *The Winter Soldiers: The Battles of Trenton and Princeton* (Doubleday Publications, 1973), 291.
2. John T. Cunningham, *The Uncertain Revolution - Washington & the Continental Army at Morristown* (West Creek, N.J., Cormorant Publishing, 2007), 18-44.
3. Sherman, Andrew M., "Washington's Army in Lowantica Valley, Morris County, New Jersey Winter of 1776-1777," *American Historical Magazine* (Volume III January, 1908).
3. Richard Ketchum, *The Winter Soldiers: The Battles of Trenton and Princeton* (Doubleday Publications, 1973), 291.
4. McCoy, Eric A. Major, *Army Sustainment, The Impact of Logistics on the British Defeat in the Revolutionary War*, 700-12-05 volume 44, Issue 5. September -October. 2012.

Chapter 2. Battle of Millstone [pages 8-24]

1. Fitzpatrick, John C. *The Writings of Washington from Original Manuscript sources, 1745-1799.*
2. Stedman, Charles, *"History of the Origin, Progress and Termination of the American War,"* in R. Kent Newmyer's *The American Historical Review* Vol. 63, No. 4 (July, 1958), 924-934
3. Ewald, Johann, *Diary of the American War, A Hessian Journal.* Translated and edited by Joseph Tostin (New Haven: Yale University Press, , 1979), 51.
4 Rodney, Thomas, "Diary" Wilmington, Delaware Historical Society Papers, VIII, 1888, 32-38.
5. Letter is cited by Glenn and Georgeanne Valis, source "Rutgers Library." The Battle of Millstone, is at http://www.doublegv.com/ggv/battles/millstone.html ("Valis") (accessed 4 October 2018).
6. Robertson, Archibald, *His Diaries and Sketches in America 1762-1780* (New York, Lydenberg, Harry, ed., reprinted New York Public Library, 1971), 122.

Endnotes 337

7. Lossing, Benson John, *The Pictorial Fieldbook of the Revolution* (New York: Harper and Brothers, 1859), Vol.1. Ch.XVI, 331.
8. The account of Samuel Sutphen, slave and militia soldier is quoted from *New Jersey in the Revolution, a Documentary History*, edited by Larry Gerlach, (Trenton: N.J. Historical Commission, 1975), 354 – 360.
9. Houston William Churchill, Journal (the *Princeton Standard*, May 1, 1863), as quoted from *New Jersey in the Revolution, a Documentary History*, 334-336.
10. This is from a typescript of Sullivan's journal, 1775-1778. The original is at the American Philosophical Society in Philadelphia, The journal was published in 1997 as *From Redcoat to Rebel: The Thomas Sullivan Journal*, by Joseph Boyle. Thomas Sullivan was born in Ireland and arrived in American in June 1775. He served as a sergeant in the 49th Regiment of Foot from the siege of Boston in 1775 through the last days of the Philadelphia Campaign in 1778. He deserted the British Army in June, 1778, and subsequently became the steward of General Nathanael Greene.
11. Library of Congress, Map of British Outposts between Burlington and New Bridge, New Jersey, December 1776, online at http://www.loc.gov/item/gm71002189/ (accessed 6 October 2014). Another interesting map of the area is maps: http://www.westjerseyhistory.org/maps/revwarmaps/hessianmaps/index2.shtml (accessed 8 October 2018).
12. Ewald, *Diary of the American War*, 59.

Chapter 3. The Forage Wars [pages 25-34]

1. Washington to Hancock, January 7, 1777. *Washington's Papers*.
2. Ferling, John, *Almost a Miracle*, (Oxford: Oxford University Press, 2007), 205. Estimates indicate that the Continental Army was reduced to about 800 soldiers by the end of January 1777.
3. New Brunswick is consistently referred to as "Brunswick" in contemporary documents; it typically encompasses adjacent towns of today's Franklin Township, Somerset and Milltown.
4. Sullivan, Thomas, *Journal of the Operations of the American War 1775-1782*. A typescript of the Sullivan Journal is available at the Historical Society of Pennsylvania (Collection 1098).
5. *Pennsylvania Evening Post*, February 6, 1777.
6. Ewald, *Diary of the American War*, 53
7. Dickinson to Washington, February 9, 1777, *Washington's Papers*.

8. Ward, Henry M., *General William Maxwell and the New Jersey Continentals*, Military Studies Book 168 (Santa Barbara, CA: Preager, 1997), 57.
9. Peebles, John, *John Peebles' American War, Journal of Captain John Peebles, 42nd Regiment of Foot, attached to the 1st Battalion Grenadiers, 21 September, 1776-15 October 1778.* (Publications of the Army Records Society, Vol. 13).
10. Charles Stedman, *The History of the Origin, Progress, and Termination of the American War*, (London: 1794), 1:241.
11. Lydenberg, Harry Miller, ed., *Archibald Robertson, Lieutenant-General Royal Engineers: His Diaries and Sketches in America, 1762–1780* (New York: New York Public Library, 1930), 122.
12. Charles Stuart to Lord Bute, March 19, 1777 in *A Prime Minister and His Son*, Mrs. E Stuart Wortley, ed. (London, 1905), 103.

Chapter 4. The Bloody Battle of Bound Brook [pages 35-48]

1. Washington to John Hancock, January 7, 1777, *Washington's Papers*, VIII.
2. Mattern, David B., *Benjamin Lincoln and the American Revolution*, (Columbia: University of South Carolina Press, 1992), 22.
3. Swan, Kells H., Unpublished Manuscript, March 18, 2003
4. Ewald,. *Diary of the American War*, 55.
5. Ibid., 56.
6. Ibid., 57.
7. Ibid., 57.
8. Ibid., 58.
9. Atwood Rodney, *The Hessians* (Cambridge, 1980) 117.
10. Ewald, *Diary of the American War*, 55
11. Ibid., 57.
12. Davis, Rev. T. E., *The Battle of Bound Brook: An Address Delivered Before the Washington Camp Ground Association*, (Bound Brook, NJ: The Chronicle Stream Printery, 1894), 6.
13. Ibid., 10.
14. McGuire, Thomas J., *The Philadelphia Campaign, Vol. I: Brandywine and the Fall of Philadelphia*, (Mechanicsburg, PA. Stackpole Books, 2006.), 22.
15. Nead, Benjamin M. . "A Sketch of Gen. Thomas Proctor, with some account of the First Pennsylvania Artillery in the Revolution". (*Pennsylvania Magazine of History & Biography*. Vol. 4 No. 4. 1880)
16. McGuire, *The Philadelphia Campaign*, 23.

Endnotes

17. Ward, Christopher, *The War of the Revolution,* (New York: MacMillan, 1952), 27-37.

Chapter 5. The First Middlebrook Encampment [pages 49-64]

1. Muhlenberg, Henry A. *The Life of Major General Peter Muhlenburg* (Philadelphia, Carney and Hart, 1849), 76. Also Magill, Jacob, *Somerset County Historical Society Quarterly II,* 1912, 105.
2. Magill, Jacob, *Somerset County Historical Society Quarterly II,* 1912, 105.
3. Ibid., 71
4. Spears, John B., *Anthony Wayne,* (New York: D. Appleton & Co., 1903), 61-62.
5. Fitzpatrick VIII, General Orders June 2, 1777, 70.
6. Ibid., 71.
7. Muenchhausen, Friedrich, *At General Howe's Side, 1776-1778: The Diary of General William Howe's Aide-de-Camp,* (Monmouth Beach, NJ: Philip Freneau Press, 1974), 14.
8. Washington to Joseph Reed, June 23, 1777, Volume 10, *Washington's Papers.*
9. Fitzpatrick VIII, General Orders May 22, 1777, 99.
10. Scull, W., *Road from Quibbletown to Amboy,* (Erskine DeWitt Series No. 55], 1779. Scale 1 inch: 2 miles.
11. Andre, J. [probable author]. Untitled - *Parts of the Modern Counties of Middlesex and Somerset 1779.* Scale 1 inch: 2.5 miles.
12. Somerset County Tavern Licensees 1776-1806.
13. Fitzpatrick VIII, General Orders June 6, 1777. 72.
14. Messler, Abraham D. D., *History of Somerset County* (Somerville, NJ: C. M. Jameson Publisher, 1878), 83. Extracted from "Official Letters of General Washington to the American Congress" Vol. ii. Boston, 1796, 123-124.
15. Fitzpatrick, Washington to President of Congress, May 5, 1777.
16. General Orders, Middlebrook, June 11, 1777, 73.
17. *Sullivan's Papers* I, 350, 360-365, 375-376.
18. Greene's Papers, Greene to Lincoln April 19 1777.
19. Washington to John Augustine Washington, from "Camp at Middlebrook," June 29, 1779.
20. Muenchhausen, *At General Howe's Side,* 16.
21. Wildes, Henry Emerson, *Anthony Wayne: Troubleshooter of the American Revolution,* (New York: Harcourt, Brace & Company, 1941), 106.

Chapter 6. The Nest of the American Eagle [pages 65-77]

1. John Adams to Abigail Adams, 14 April 1776, Founders Online, National Archives.
2. Thacher, James, M.D., *A Military Journal During the American Revolutionary War, From 1775 to 1783* (New York, Arno Press, 1969), 152.
3. *Somerset County Historical Quarterly*, Vol 1. (Somerville, N.J. January, 1912) 14.
4. Washington to Joseph Reed, Fitzpatrick X. June 23, 1777.
5. Muenchhausen, Friedrich Ernst von, *At General Howe's Side, Muenchhausen's Diary* (Philip Freneau Press, 1974)16.
6. George Washington to John Hancock, June 20, 1777. NARA, Founders Outline, www.Founders.archives.gov.
7. The Diary of Quartermaster Sergeant Simon Griffin of Col. Samuel B. Webb's Regiment 1777-1779; Microfilm and bound copies of the originals at the Connecticut State Library and Archives in Hartford.
8. Andre, John, Major Andre's Journal. Stirling's Brigade rejoined Washington's troop at Middlebrook.
9. Washington to Stirling, Fitzpatrick, X, June 23, 1777.
10. The original Drake farmhouse has been preserved and is today a public museum administered by the Historical Society of Plainfield.
11. Peale, Charles Willson, *The Selected Papers of Charles Willson Peale* (Yale University Press, 1988 vol. 1), 236.

Chapter 7. The Battle of the Short Hills [pages 78-96]

1. Ruppert, Robert, "The First Fight of Ferguson's Rifle," *Journal of the American Revolution*, November, 2014.
2. New Jersey Archives, State Library, Trenton N.J., Essex County Inventories, 86.
3. Messler, Abraham, D.D. *Centennial History of Somerset County* (Somerville, N.J., C.M. Jamison, 1878), 91.
4. Charles Stuart to Lord Bute, "New York, July 10, 1777," New Records, 33. Also in McGuire, Thomas, *The Philadelphia Campaign, Volume 1: Brandywine and the Fall of Philadelphia* (Mechanicsburg, PA, Stackpole, 2006) Chp. 1.

Chapter 8. The Rampaging Retreat [pages 97-106]

1. Detwiller, Frederic C., *War in The Countryside, The Battle and Plunder*

of the Short Hills New Jersey, June 1777 (Plainfield, NJ, Interstate Printing Corp.), Chapters V,VII, VIII.
2. *The Pennsylvania Evening Post*, July 10, 1777.
3. New Jersey Archives, State Library, Trenton, N.J. The extensive plundering of the countryside by both British and American troops before and after the battle led to a large number of damage claims from civilians. An exact description of the property and often the dates when the damages occurred appears in these inventories. Several are dated the week of the battle. An examination of these records provides a representative picture of the contents of the farmhouses and shops and craftsmen's tool of the era.
4. *Collections of New Jersey Historical Society*, vol. VI. 99.
5. Ricord, Frederick W., *History of Union County*, 503.
6. Peter Wilson, comp., *Acts of the Council and General Assembly of the State of New-Jersey from the Establishment of the Present Government, and Declaration of Independence, to the End of the first Sitting of the Eight Session, on the 14th Day of December, 1783*(Trenton: Isaac Collins, 1784), 237.
7. New Jersey Revolutionary War Damage Claims, Claims Against the British, Essex & Middlesex Counties, Reel 2, New Jersey State Archives. A total of 115 individuals filed claims for damages in Westfield during the entire war, with 92 claiming damages for June 26-27, 1777.
8. Andre, *Andre's Journal*, 429.
9. Ibid., 33
10. Israel Shreve to Dr. Otto Bodo, June 29, 1777, in John U. Rees, *Colonel Israel Shreve's Journal, 23 November 17 1776 to 14 August 1777*. http://www.scribd.com/doc/153790118 (accessed October 3, 2017).
11. Alexander Hamilton to William Livingston June 28, 1777, New York Historical Society, Museum Collection, Letter was addressed to Livingston while he was a member of the Committee of Correspondence.
12. Muenchhausen, *At General Howe's Side*, 17.
13. Detwiller,*War in the Countryside*, 1.
14. Stillman,George B., "Battle of the Short Hills. Reenactment preparations June 29 and 30th 2002" *The Brigade Courier, Brigade of the American Revolution*, Volume 18, 3.
15. Composite of Robert Erskine Maps of the Short Hillsand Ash Swamp. C.1779, No. 74A, 74B, 78. Collections of the New York Historical Society. Major Robert Erskind was commissioned as the official surveyor for the American Army in 1777 and produced an extensive collection.

Chapter 9. Why Middlebrook? [pages 107-115]

1. Fitzpatrick, *Washington's Papers*, XIII, 75-76.
2. General Orders, February 5, 1783.
3. Fitzpatrick, Greene to Washington, 18 Oct. 1778; Washington to Stirling, 24 Oct. 1778, *Washington's Papers*.
4. Lord Stirling. to Washington Oct 26, 1778, *Washington's Papers*, LXXXIX, 123.

Chapter 10. The American Army Arrives [pages 116-126]

1. Fitzpatrick, *Washington's Papers*, VI, 134.
2. Angelakos, Peter, "The Army at Middlebrook," *Proceedings of the New Jersey Historical Society* LXX (1952).
3. Lossing Vol. 1. 33.
4. Baron DeKalb to Washington, February 8, 1779, *Washington's Papers*, XCVII, 112.
5. Erskine Robert, Road from New Brunswick to Bound Brook, Map No. 70E. The original Erskine maps are housed at the New York Historical Society.
6. Fitzpatrick, November 1, 1778-14 January 1779, Vol. 18, 447-452.

Chapter 11. Settling In [pages 127-133]

1. *Washington's Papers*, XIII, 80
2. Green to Pettit, December 4, 1778 , *Greene's Papers*.
3. *Washington's Papers*, XIII, 453.
4. Fisher, Elijah, *Journal While in the War of Independence*. (Augusta:, Badger and Manley, 1880), 11.
5. Wayne to Joseph Reed December 28 1778, *Wayne Papers*, II, 197.
6. Thacher, Chapter 5, 158 James, M.D., A Military Journal During the American Revolutionary War, From 1775 to 1783, (New York: Arno Press, 1969).
7. Stirling to Washington, December 28, 1778, *Washington's Papers*, XCV, 34.
8. Brigade Orders January 27, 1779 Smallwood Orderly Book.
9. Chatellux, Francios, *Travels in America 1780-81-82 (NewYork, 1828)*, 802.
10. Thacher, 106.
11. Weiss, Jacob, *The Letter Book of Deputy Quartermaster General of the*

Endnotes 343

Revolution. Melville J. Boyer ed. (Allentown, 1956. Lehigh County Historical Society, Volume 21.), 118.
12. Fitzpatrick, Washington to Lafayette, 8–10 March 1779, vol. 19, 15 January–7 April 1779.

Chapter 12. Everyday Life, [pages 134-145]

1. Danckert, Stephen C., "Baron von Steuben and the Training of Armies,"(*Military Review* 74, 1994), 29-34.
2. Orderly Book , April 6, 1776, Lewis, Andrew, General Orderly book.
3. Richard Clairborne to James Abeel April 25, 1779, Abeel Papers at New Jersey Historical Society.
4. Furneaux, Rupert, *The Pictorial History of the American Revolution as told by Eye Witnesses and Participants* (Lansing: University of Michigan, J.G. Ferguson, 1973), 108.
5. Service Record, Pride, James, W 26939, U. S. National Archives.
6. Fisher, *Military Journal*, 186.
7 Wild, Ebenezer, *Journal of Ebenezer Wild*, 1779, (Boston, MA: *Proceedings of the Massachusetts Historical Society*, 2nd series, vol. VI 1891), David Library Collection, 79-160.
8. Whiting, John, *Revolutionary Orders Of General Washington: Issued During The Years 1778, 1780, 1781, And 1782* (1844; repr., Whitefish, Montana: Kessinger Publishing , 2010), 75.
9. Journals of the Continental Congress Vol II, 112.
10. Coits Orderly Book (June 14, 1775), 19.
11. Gardner, "Last Cantonment of the Continental Army" (*American History Magazine* Vol X), 369.
12. Bolton, Charles K., *The Private Soldier Under Washington* (New York: Kennikat Press, 1902), 107, 161.
13. Gano, John, "Rev. John Gano's Biographical Memoirs," (New York: 1806). Also in *New York Historical Magazine*, Vol 5, 332.

Chapter 13. The Women of Middlebrook [pages 146-155]

1. *Washington's Papers*, VIII, 181. General orders, June 4, 1777.
2. John U. Rees, "The multitude of women: An Examination of the Numbers

of Female Camp Followers with the Continental Army," *The Brigade Dispatch*, vols. xxiii to xxviii, from Records, 1775-1790's, Record Group 93, National Archives Microfilm Publication M859, reel 76, items no. 22185, 22186, 22187, 22188 and 22189.
3. "Revolutionary Services of Captain John Markland," *Pennsylvania Magazine of History and Biography*, vol. 9 (1885), 105.
4. Washington to the Superintendent of Finance, 29 January 1783, *Washington's Papers*, XXVI, 7879.
5. Regimental orders, September 30, October 7, 1778, Orderly Book of the 2nd Pennsylvania Regiment, 1778,. Regimental orders, August 14, 1782, Orderly Book of the 10th Mass. Regt., 1782, Daughters of the American Revolution Museum.
6. *Washington's Papers*, XIV, 422, 423.
7. Pension and Bounty Land Warrant Application Files, Patrick Cronkite, Fifer, 1st New York Regiment, 1777-1783, W16932, contains supplementary depositions of Maria Cronkite (nee Humphrey). U. S. National Archives.
8. Lesser, Charles H., ed., "Return of Negroes in the Army, 24 August, 1778," *Brigade Dispatch*, Vol XXVIII, No. 1, Spring 1998, 9.
9. Greene to Washington, June 7, 1779, *Greene's Papers*, VI, 31.
10. Blumenthal, Walter H., *Women Camp Followers of the American Revolution* (Philadelphia: George Mac Manus Co., 1952), 65, 66.

Chapter 14. Festive Events and Fascinating Visitors [pages 156-166]

1. *Washington's Papers*, XIV, 455.
2. Girard to Washington, May 5, 1779. Fitzpatrick CVI, 17.
3. Wright, Robert K. and MacGregor, Morris J. *"William Livingston, Soldier-Statesmen of the Constitution*. (Washington, DC: United States Army Center of Military History. CMH Pub.), 71-25.
4. Benedict Arnold to Peggy Shippen, Febury 8, 1779, *Joseph Reed Papers* VI, New York Historical Society.
5. Puls, Mark, *Henry Knox Visionary General of the American Revolution* (Laurenburg, N.C.: St Andrews Press, 2008), 13.
6. *Washington's Papers*, December 17, 1777.
7. Fisher, Elijah. *Journal While in the War for Independence, 1775-1784*, (Augusta, Maine: Badger and Manley, 1880).
8. Thacher, 157.
9. Ibid., 157

Endnotes 345

10. Major Forsythe to Greene, January 27, 1778, *Greene's Papers* VIII. 25.
11. Jedutha Baldwin, Colonel, *Revolutionary War Journal 1775-1778*, (Sagwan Press, 2018). Original is housed at the David Library of the Revolution.
12. *Greene's Papers*, Vol. 3.

Chapter 15. Crime and Punishment [pages 167-175]

1. Stirling to Abeel, Jan. 29, 1779. *Stirling Papers*, IV, 98.
2. *Washington's Papers* Vol. XV, 207
3. Ibid., XIV, 424-426.
4. Ibid., 376-377.
5. *Washington's Papers*, XIV, 425.
6. Cox, Caroline, *A Proper Sense of Honor: Service & Sacrifice in George Washington's Army*, (Chapel Hill: University of North Carolina Press, 2004), 94-98.
7. *Washington's Papers*, Washington to Colonel Goose Van Schaick, Oct.27, 1778; Washington to Brigadier General Samuel Holden Parsons, April 25, 1777; Washington to Brigadier General George Clinton, May 5, 1777.
8. Thacher, James, *Military Journal*, 252.
9. Ibid., 196.
10. *Deseret News and Telegram*, Salt Lake City, April 9, 1962.
11. General Orders, Sept. 16, 1775, Jan. 5, 1778, Feb. 8, 1778, March 14, 1778, May 11, 1783.
12. General Orders, Jan. 1, 1776, June 10, 1777.

Chapter 16. Supporting a Deprived Army [pages 176-190]

1. Erskine Map No. 70F.
2. Erskine Map No. 70A.
3. "The Army at Middlebrook, 1778–1779," *Proceedings of the New Jersey Historical Society* (April 1952), 97–119.
4. *Washington's Paper's*, II, 967.
5. Ibid., XIII, 472.
6. Ibid, II, 975-976.
7. Bolton Charles Knowles, *The Private Soldier Under Washington*, (New York: Scribner's, 1902), 78.
8. Brigade Orders March 6, 1779, Smallwood's Orderly Book.

9. Expenses from May 1778 to January 1779, Azariah Dunham Account Book Dunham papers, *Revolutionary War Manuscripts in Special Collections and Archives, Rutgers University Libraries.*
10. Return of Horses, May 1779, *Greene's Papers,* IX, 28.
11. Major Storey to Greene, May 20, 1779, *Greene's Papers* V, 37.
12. Greene to Nehemiah Hubbard, *Greene's Papers,* January 27, 1779.
13. Greene to Nehemiah Hubbard, *Greene's Papers,* January 27, 1779.
14. *Washington's Papers,* XIV, 412.
15. Martin Joseph Plumb, *A Narrative of a Revolutionary War Soldier,* (New York: Signet, 2001), 208.
16. Christopher Hibbert *Redcoats and Rebels,* (New York: Norton Books, 1990), 257.
17. Hooper to Greene, *Greene's Papers,* February. 18, 1779, IV 42.
18. Risch, Emma, *Quartermaster Support of the Army,* 46.
19. *Washington's Writings* XCV, Stirling to Washington, Dec. 30, 1778.
20. Abeel to unknown, December 5, 1778, Abeel letter book, 142. Housed at the New Jersey Historical Society.
21. Biddle to Greene May 27, 28, 1779. *Greene's Papers* VII, 52.
22. Mayers, Robert A., *Searching For Yankee Doodle,* 96.
23. Sekel, Clifford, *The Continental Artillery in Winter Encampment at Pluckemin, New Jersey* (Wagner College, 1972) 55.
24. Rau, Louise, "Sergeant John Smith's Diary of 1776," *The Mississippi Valley Historical Review,* Vol. 20. No. 2 (Sept. 1933), 247-270.
25. Baldwin, Jeduthan, Baldwin, Thomas W. ed., *Revolutionary Journal 1778-1779,* (Bangor: 1906), 142.

Chapter 17. General Greene's Worst Nightmare [pages 191-198]

1. Greene to Washington, April 24, 1779, *Greene's Papers,* III, 426-427.
2. Greene to McDougall quoted in an editor's note at *Greene's Papers,* II, 308-310.
3. Risch, Erna, *Quartermaster Support of the Army,* (Special Publications CMH Pub 70-35, 1962, 1989), 41, 43-45.
4. Furman to Greene, May 19, 1779, *Greene's Papers* II, 228.
5. Wayne to John Van Nest January 16, 1779, *Anthony Wayne Papers,* VI.
6. Stirling's Journal, Dec 3, 1778 , *Stirling Papers,* IV, 97
7. Thacher, *Military Journal,* 161.
8. *St. Clair Papers,* I, 465.

Endnotes 347

9. *Washington's Papers*, XIV, 56.
10. Washington to Greene, August 15, 1780, *Greene's Papers*, VI, 17.

Chapter 18. Civilian Support [pages 199-206]

1. *Washington's Papers*, XIV, 412.
2. *Washington's Paper's*, XIII, 453.
3. Scammell to Wayne, January 14, 1775, *Wayne Papers*, VI, 57.
4. Brigade Orders, March 6, 1779, Smallwood Orderly Book.
5. Greene to Wayne, December 19, 1778, *Wayne Papers*, VI, 39.
6. Greene to Jeremiah Wadsworth, April 14, 1779, Bancroft Collection, II, 297. (New York Public Library).
7. Unknown person to Governor Franklin. February 3, 1779, Bancroft Collection. I, 93. (New York Public Library).
8. *Washington's Papers*, XV, 210.
9. Dr. Binney to Major Story, May 19, 1779, *Greene's Papers*, V, 42.

Chapter 19. The Army Departs [pages 207-219]

1. *Somerset County Historical Quarterly*, (Somerville, NJ, January 1912, Vol 1, No. 1,) 6.
2. Maas, E. J. ed., *North of the Rariton Lotts*. (Neshanic Printing Company, Neshanic, New Jersey, 1975), 135-136.
3. *Washington's Papers*, XV, 74.
4. *Washington's Papers*, VI, 134, 267-268.
5. Boom, A.A. "Report on Middlebrook Encampment by the Continental Army during the Middle of 1777 and the Winter of 1778-79". *Somerset County Historical Quarterly*, 1975, Somerville, New Jersey.
6. Simcoe, John A., *Journal of the Operations of the Queen's Rangers, from the End of the Year 1777, to the Conclusion of the Late American War*, (Exeter, UK, printed by the author, 1787).
7. Simcoe, John G., *Simcoe's Military Journal: A History of the Operations of a Partisan Corps, Called the Queen's Rangers* (New York: Bartlett & Wilford, 1844), 325.
8. Ibid., 144-145
9. McPike, Eugene F., "The Capture of Lieutenant Colonel J.G. Simcoe - An incident of the American Revolution", *The American Monthly Magazine*,

Vol. XI (Washington, D.C.: National Society, D.A.R., 1897), 564.
10. *The Washington-Rochambeau Revolutionary Route in the State of New Jersey, 1781—1783. An Historical and Architectural Survey*. 3 Vols. (Trenton, NJ: New Jersey Historic Trust, Department of Community Affairs, 2006).
11. Orderly Book 2nd New York Regiment for August 29 1781.

Chapter 20. "Rank Hath Its Privileges" [pages 220-231]

1. Beatty to Greene Feb. 17, 1779, Greene's Papers, IV, 5.
2. Washington's Writings, Livingston to Washington, 9 March 1779; Washington to Livingston, 23 March, 1779; Washington to Anthony Wayne, 16 March 1779; and Wayne to Washington, 23 March 1779.
3. Revolutionary War Accounts, Vouchers, and Receipted Accounts 1, 1776–1780, DLC:G W, Ser. 5.
4. Williams, Glen F., *The Year of the Hangman* (Yardley, PA: Westholme Publishing, 2005), 214.
5. Davis, T. E. *The Battle of Bound Brook* (Washington Campground Association, Appendix A, 1895), 25–26.
6. Von Steuben to M. Du Ponceu, May 3, 1779, *Joseph Reed Papers*, vol VI.
7. Melick, Arthur D., *The Story of an Old Farm, or Life in New Jersey in the Eighteenth Century, with a Genealogical Appendix*(Someville, N.J.: *Unionist Gazette*, 1889), 53.
8. General Nathanael Greene to Colonel Jeremiah Wadsworth, Middle Brook, New Jersey, March 19, 1779.
9. Ibid., Davis, 16.
10. Sparks, Jared, ed., *Correspondence of the American Revolution* (Boston 1853, vol. II), 266-268.

Chapter 21. Washington on the Rocks [pages 232-248]

1. Andrews, John, L.L.D., *A History of the War with America, France, Spain, and Holland. Commencing in 1775 and Ending in 1783.* (London, 1785 Printed for John Fielding, and John Jarvis, 1785, by the King's Royal License and Authority).
2. Messler, Abraham, *Centennial History of Somerset County* (Somerville: C. M. Jameson, 1878), 83.
3. In 1973-74, A. A. Boom, a local historian, conducted tours of the area

Endnotes

around Miller Lane and did an extensive investigation of the rock walls in the area, as documented in Chapter 11 of *"North of the Rariton Lotts: A history of Martinsville, NJ Area,"* Edward J. Maas, Editor, Martinsville Historical Committee, 1975. In 1974, William Liesenbein performed an archaeological investigation of the same area and refutes Boom's claims that most of the walls were set up for fortifications, but did admit that some may have been. Liesenbeim also found features which he speculated could have been huts and kitchens in 1777.

4. Lossing Vol. 1, Chp. XVI, 33
5. Map number 55 in the Erskine/DeWitt series, was drawn for Capt. William Scull. It shows detailed locations of the Continental Army units during the 1776-77 Middlebrook encampment. U.S. National Archives.
6. Wright, Robert K., *The Continental Army* (Washington, D.C.: Center of Military History, United States Army, 1986), Ch. 5, 91.
7. Barber, John Warner., *Historical collections of the State of New Jersey: Past and Present,* (New Haven, CT., J. W. Barber, 1868), 201.
8. Randolph, Edward F., *Journal of Edward Fitz Randolph* (Rutgers University Special Collections, University Archives), 6.
9. Ott, Westley H., *1 mile X 1 Mile X 100 Years Dunellen, N.J.* (Collections of the Green Brook N.J. Public Library).
10. Rebecca Vail describes a picnic trip to the Rock from Rahway in 1847 and includes a copy of a postcard showing the Rock before the stone walls were built. (Gladys Whitehead's Memory Books Recollections 2, 1986. Collections of the Plainfield, N.J. Public Library).
11. Lossing, Benson John *The Pictorial Field-Book of the Revolution, Illustrations, by Pen and Pencil, of the History, Biography, Scenery, Relics, and Traditions of the War for Independence.* (New York Harper & Brothers, 1859), Vol. 1, Chp XVI, 331.
12. John Laing named in 1880 Federal Census, City of Plainfield, N.J. First District, 52.
13. *The Constitutionist,* July 16, 1896.
14. Ibid., July 14, 1898.
15. Ibid., February 27, 1868.
16. Bebbington, George and Siegel, Alan A., "Washington Rock Is Focus Of Renewed Interest," (*Warren History,* Volume Two, No. 3, Spring 1995 Warren Township Historical Society).

Chapter 22. Scarce Research [pages 249-261]

1. Maas Edward J., ed., *North of the Rariton Lotts-A History of the Martinsville, New Jersey area* (Neshanic Station, N.J.: Neshanic Printing Co., Inc. 1975).
2. Ibid., Mass, 103.
3. Ibid., Mass, 105.
4. Liesenbein, W. *Report on a Preliminary Archaeological Investigation of the Alleged Site of the 1777. Summer Encampment of Wayne's Brigade at Middlebrook, New Jersey, June 9 to July 27, 1974*. (Report on file, Department of History, Rutgers University), 23.
5. Ibid., Liesenbein, 27.
6. Kardas, S., and E. Larrabee 1977, "Cultural Resource Survey for Prehistoric Sites, Historic Sites and Structures Along the Bridgewater Collector Systems." On file, New Jersey Historic Preservation Office (NJDEP), Trenton, New Jersey.
7. Rutsch, E. S., and R. L. Porter, 1981-82 draft report, "Research Materials and Field Documentation from an Archaeological Survey for the Elizabethtown Water Company Water." Unpublished manuscript materials on file, New Jersey Historic Preservation Office (NJDEP), Trenton, New Jersey.
8. Bower, Ernest. "Revolutionary War at the Crossroads." *The Treasure Depot Magazine* 3 (June 2000).
9. *A Historical and Archeological Study of Washington Valley Park, East of Middlebrook Including Part of the Site of Wayne's Brigade Encampment of 1777*. (Bridgewater Township, Somerset County N.J. Prepared for the Somerset County Park Commission by Hunter Resource Consultants, Trenton, N.J. 2003).
10. Messler, 83.
11. Davis, 28.
12. Angelakos, Peter "The Army at Middlebrook 1778-1779," (*Proceedings of the New Jersey Historical Society*, vol. 70, no. 2, 1952).
13. Prince, 1-87.
14. *Washington's Papers*, XIV, 182.
15. The Erskine-DeWitt map series is itemized in *American Maps and Map*
16. Guthorn, Peter J., *Makers of the Revolution*, (Monmouth Beach, New Jersey: Philip Freneau Press, 1966).

Endnotes

Chapter 23. With Knox at the Artillery Camp [pages 262-274]

1. *Knox Papers*, III, at Massachusetts Historic Society.
2. *Washington's Papers*, Xlll, 418.
3. Colonel John Lamb's Second Regiment of Continental Artillery, Orderbooks and book of monthly and weekly returns 1778-1783 On microfilm at the New York Historical Society and David Library, Philadelphia.
4. Excavations at Knox's Artillery Encampment, Pluckemin. Series of articles in the *Bernardsville News*, Bernardsville, New Jersey, 1917.
5. Parry, Samuel "The Origin of the Name 'Pluckemin.'" *Somerset County Historical Quarterly*, 1. 196, (1912).
6. Sekel, Clifford, *"The Continental Artillery in Winter Encampment at Pluckemin, New Jersey, December 1778-June 1779,"* (Master of Arts Thesis, Wagner College, 1972), 16.
7. Seidel, John Lewis, "The Archeology of the American Revolution: A Reappraisal and Case Study of the Continental Artillery Cantonment of 1777-1." (To the faculties of the Universty of Pennsylvania, in Partial Fulfillment of the Requirements for the Degree of Doctor of Philosophy).

Chapter 24. A Well-Built Surprise [pages 275-285]

1. Regimental Orders December 9, 1778, Orderly book 2.
2. Ibid., December 26, 1778.
3. Knox to McDougal, January 10, 1779, *McDougal Papers*, New York Historical Society.
4. Regimental Orders, December 6, 1778. Col Lamb's Orderly Book.
5. Knox to McDougal, January 10 1779, Alexander McDougal Papers.
6. Regimental Orders, February 2, 1779, Lamb's Orderly Book.
7. Regimental Orders, December 8, 1778, Lamb's Orderly Book.
8. Regimental Orders, February 23, 1779, Lamb's Orderly Book.
9. The personal observations and examination of the site by Clifford Sekel, correlated with articles by Schrabisch in the Bernardsville News from 1917.
10. Moore, *Diary of the American Revolution*. II, 132.
11. Regimental Orders, May 22, 1779, Lamb's Orderly Book.
12. Thacher, Journal, 53.
13. Regimental Orders, February 15, 1779, Lamb's Orderly Book.
14. *Washington's Writings*, XV, 49.
15. Ibid., 19:577.

16. Regimental Orders, December 14, 1779, Lamb's Orderly Book.
17. Havens, Jessie, "Beacon Network, Middlebrook and Morristown Encampments", map and description, Township of Montgomery Geographic Information System, based on data from J. T. Raleigh, Philip Freneau Press, Jan. 2004.

Chapter 25. Our Compassionate General [pages 286-292]

1. Brigade Orders, Lamb, February 9, 1779.
2. Battalion Orders, Lamb, February 28, 1779.
3. Brigade Orders, Lamb, February 13, 1779
4. Regimental Orders, May 21, 1779, Lamb's Orderly Book.
5. *Somerset Historic Quarterly*, VII (1917), 16.

Chapter 26. Everyday Adventures at Pluckemin [pages 293-300]

1. Brigade Orders, Lamb, February 28, 1779.
2. Orderly Book 2, Lamb February 3, 1779.
3. Regimental Orders, February 3, 1779, Lambs Orderly Book.
4. Orderly book 2, Lamb, January 9. 1779.
5. Regimental Orders, May 3 1779, Lamb's Orderly Book.
6. Sekel, 53.
7. Hazard , ed., Pennsylvania Archives VII, 121.
8. Ibid., 254.
9. Knox to Stirling, December 27, 1778, *Knox Papers*, VIII. Prince, 27.
10. *Washington's Papers*, Vol XV 220-221.
11. Knox to McDougall Jauuary 10, 1779, *McDougall Papers*.
12. *Washington's Writings*, XV 220. 221.

Chapter 27. Pluckemin Rediscovered Then Lost Again [pages 301-312]

1. Seidel, 683.
2. *Letters Sent by Samuel Hodgdon, 07/1778 – 05/1784. Supply Records Numbered Record Books.*National Archives Microfilm Publication M853, reel 33, vol. III, no. 127-8.
3. Veit, Richard, F. *Digging New Jersey's Past: Historical Arceology in the*

Endnotes 353

Garden State (Rutgers University Press, 2002), 68-71.
4. Schrabisch, Henry, *Excavations at Knox's Artillery Encampment, Pluckemin*. Series of articles in the *Bernardsville News*, Bernardsville, New Jersey, 1917.
5. "General Knox's Artillery Park, Pluckemin," *Somerset County Historical Quarterly*, 6 (3): 161-168.
6. Seidel, John Lewis, *The Archeology Of The American Revolution: A Reappraisal and Case Study of The Continental Artillery Cantonment of 1778-1779*, University of Pennsylvania, in Partial Fulfillment of the Requirements for the Degree of Doctor of Philosophy, 1987.
7. Seidel, John L. (1983) "Archaeological Research at the 1778-79 Winter Cantonment of the Continental Artillery, Pluckemin, New Jersey," *Northeast Historical Archaeology*: Vol. 12, Article 4.

Chapter 28. Conclusion and Epilogue [pages 313-319]

1. Martin, Joseph Plumb, *Private Yankee Doodle* (New York: Little Brown, 1962), 78.
2. *WPA Guide to 1930s New Jersey*, (Viking Press 1939, reprinted by Rutgers University Press, 1986), 431.

Bibliography

Abeel, Colonel James, Letter and Receipt Book for 1778-1779. MS. in Morristown Historic Park Museum.

Anderson, Troyer S. *The Command of the Howe Brothers during the American Revolution.* (New York: Oxford University Press, 1936).

Andre, John, *Major Andre's Journal, Operations of the British Army, June 1777 to November 1778.* (Tarrytown, NY: William Abbatt, 1930; reprinted New York: The New York Times & Arno Press, 1968).

Angelakos, Peter, "The Army at Middlebrook 1778-1779." *Proceedings of the NJ Historic Society.* (LXX, No. 2: 2001).

Atwood Rodney, *Mercenaries from Hessen-Kassel in the American Revolution.* (Cambridge University Press. 2002).

Baker, Norman, *Government Contractors: The British Treasury and War Supplies, 1775-1783.* (London: Athlone Press, 1971).

Bedminster Township Ratables 1778-1779. (Trenton, NJ: New Jersey State Archives).

Bell, J.B., *The Myth of the Guerilla: Revolutionary Theory and Malpractice.* (New York: Alfred A. Knopf, 1971).

Bill, A.H. *New Jersey in the Revolutionary War.* New Brunswick, NJ: Rutgers University Press, 1964).

Bland, Humphrey, *A Treatise on Military Discipline.* (London: D. Minwinter, P. Knapton, 1762).

Boom, A.A. "North of the Rariton Lotts: A History of Martinsville, NJ Area. Report on Middlebrook Encampment by the Continental Army During the Middle of 1777 and the Winter of 1778-79." (Neshanic, New Jersey: Neshanic Printing Co. 1975).

Bowen, F. "Life of Benjamin Lincoln." In *The Library of American Biography.* (Boston: Second Series. Charles C. and James Brown, 1847).

Bower, E.R. "At the Crossroads." *The Treasure Depot Magazine* 3 (June 2000).

Bradford, S. (ed.) "A British Officer's Revolutionary War Journal 1776-1778". *Maryland Historical Magazine* 56(2) 1961.

Bridgewater Township Tax Rateables, 1773-1822. (Trenton, NJ: New Jersey State Archives).

Bibliography 355

Brooks, Noah,1900 *Henry Knox, A Soldier of the Revolution*. (New York: G.P. Putnam and Sons, 1900).

Boatner, Mark M., *Encylopedia of the American Revolution*. (New York: David McKay, 1966).

Bolton, Charles, *The Private Soldier Under Washington*. (1902, Reprinted Port Washington, NY: Kennikat Press, 1964).

Callahan, North, *Henry Knox, George Washington's General*. (New York: Rhinehart,1958).

Carp, Wayne, *To starve the Army at Pleasure, Army Administration and the American Political Culture 1775-1783*. (Chapel Hill: University of North Carolina Press, 1983).

Coates, Earl J., and Kochan, James, *Don Troiani's Soldiers in America, 1754-1865*. (Mechanicsburg, PA: Stackpole Books, 1998).

Connor, M., and D.D. Scott, "Metal Detector Use in Archaeology: An Introduction." *Historical Archaeology* 32:76-85, 1998.

Collins, Isaac, 1784), New Jersey Revolutionary War Damage Claims, Claims Against the British, Essex & Middlesex Counties, Reel 2, (Trenton, NJ: New Jersey State Archives).

Cultural Resources Chapter for Washington Valley State Park Master Plan. 1996 (On file, Somerset County Park Commission, North Branch, New Jersey).

Curtis, E.E. *The Organization of the British Army in the American Revolution*. (New Haven: Yale University Press, 1926).

Davis, T.E. *The Battle of Bound Brook* an Address Delivered Beforethe Washington CampGround Association, February 22, 1894 (London, Forgotten Books, 2017).

Davis, T.E., "*The First Houses of Bound Brook.*" (Bound Brook, NJ: The Chronicle Steam Printery, 1895).

Davis, Rev. T.E., *The Battle of Bound Brook: An Address Delivered Before the Washington Camp Ground Association* (Bound Brook, NJ: The Chronicle Stream Printery, 1894).

Detwiller, Frederic C., *War In The Countryside, The Battle and Plunder of the Short Hills, New Jersey, June 1777*. (Plainfield, NJ: Interstate Printing Corp., 1976).

Dickinson, Philemon, *Papers*, (Newark, NJ: New Jersey Historical Society).

Ditchburn, Robert, "The New Jersey Encampment at Bernardsville, 1779-1780," Typescript Report on file at the Morristown Historical Park Museum, Morristown, NJ, 1971.

Dohla, Johann, Burgoyne, Bruce B.E. (ed.) *A Hessian Diary of the American Revolution.* (Norman, OK: University of Oklahoma Press, Revised ed., 1993).

Drake, A.A, Jr., R.A. Volkert, D.H. Monteverde, G.C. Herman, H. F. Houghton, R.A. Parker, and R.F. Dalton "Early Roads in Somerset County." *Somerset County Historical Quarterly* I:98-103.1912).

Dupuy, E. and N. Dupuy, *The Compact History of the Revolutionary War.* (New York: Hawthorn Books, 1963).

Ewald, Johann, *Diary of the American War, A Hessian Journal.* Joseph P. Tustin, trans. and ed., (New Haven: Yale University Press, 1979).

Ewing, Thomas, *The Military Journal of George Ewing, a soldier of Valley Forge (1754-1824).* (Yonkers, NY. Privately printed by T. Ewing, 1928).

Ferling, John, *Almost a Miracle: The American Victory in the War of Independence.* (Oxford: Oxford University Press, 2007).

Fitzpatrick. George Washington Papers. (http://memory.loc.gov/ammem/gwhtml/gwhome.html) (Washington, D. C.: Manuscript Division, Library of Congress).

Fleming, Thomas J., *The Forgotten Victory: The Battle for New Jersey, 1780.* (New York: E.P. Dutton, 1973).

Freeman, D. S., *George Washington.* (New York: Charles Scribners Sons,1951).

French, A., *The First Year of the American Revolution.* (Boston: Houghton Mifflin Company, 1934).

Gerlach, Larry H., *New Jersey in the American Revolution 1763-1783. A Documentary History.* (Trenton, NJ: New Jersey Historical Commission, 1975).

Gerlach, Larry H., *Prologue to Independence: New Jersey in the Coming of the American Revolution.* (New Brunswick: Rutgers University Press,1976).

Greene, G.W., *The Life of Nathanael Greene.* (New York: G. P. Putnam and Sons, 1867).

Greene, Nathanael, *The Papers of Nathanael Greene, October 1778.* (Chapel Hill: University Press of North Carolina, 1980).

Greiff, Constance M., *The Vanderveer House, Bedminster Township*, New Jersey, Report Prepared for New Jersey Bell Telephone. Copy on file at Pluckemin Archeological Project, 1975.

Gruber, Ira, *The Howe Brothers and the American Revolution.* (Chapel Hill: Omohundro Institute and University of North Carolina Press, 2014).

Guthorn, P.J., comp., *John Peebles' American War: The Diary of a Scottish Grenadier, 1776-1782.* (Mechanicsburg, PA: Stackpole Books, 1998).

Heitman, F. B. *Historical Register of Officers of the Continental Army During the War of the Revolution*. (Baltimore: Genealogical Publishing Co., Inc., 1982).

Hills, J., asst. engr., *A Collection of Plans in the Province of New Jersey 1777-82*. Compiled and reprinted by Peter J. Guthorn, (Brielle, NJ, Portland Press, 1976).

Hoffman, R. V., *The Revolutionary Scene in New Jersey*. (New York: The American Historical Company, 1942).

Honeyman, A. V., *Our Home: A Monthly Magazine of Original Articles*. (Somerville, NJ: Cornell & Honeyman, 1873).

Hunter, Richard, *A Historical and Archeological Study of Washington Valley Park, East of Middlebrook, Including Part of the Site of Wayne's Brigade Encampment of 1777*. (Bridgewater Township, NJ: Prepared for the Somerset County Park Commission by Hunter Resource Consultants, Trenton, NJ, 2003).

Jones, Alfred E., *The Loyalists of New Jersey: Their Memorials, Petitions, Claims, etc. From English Records*. (Newark, NJ: Collection of the New Jersey Historical Society, 1926).

Karschner, T., *Middlebrook Encampment Site. National Register of Historic Places Inventory-Nomination Form*. (Trenton, NJ: New Jersey Historic Preservation Office (NJDEP), 1973).

Karsten, Peter (ed.), *The Social Structure of the New Jersey Brigade, The Military in America*. (New York: The Free Press, 1980).

Kemble, Stephen "The Kemble Papers, Volume 1" in*Collections of the New York Historical Society for the Year 1883*. (New York: New York Historical Society, 1884).

Knox, Henry, *Henry Knox Papers*. (Boston: Massachusetts Historical Society). On microfilm at Morristown National Park, Morristown, New Jersey.

Lender, Mark E. *"The Enlisted Line: The Continental Soldiers of New Jersey"*. Ph.D. dissertation (New Brunswick: Rutgers University, 1975).

Lender, ———— *The Mind of the Rank and File: Patriotism and Motivation in the Continental Line, New Jersey in the American Revolution* III, Seventh Annual History Symposium, William Wright (ed.). (Trenton NJ, 1976).

Levitt, James H., New Jersey's Revolutionary Economy, New Jersey's, Revolutionary War Experience, 12. (Trenton, New Jersey Historical Commission, 1975).

Liesenbein, William, *Report on a Preliminary Archaeological Investigation of the Alleged Site of the 1777 Summer Encampment of Wayne's Brigade at Middlebrook, New Jersey*. Department of History (New Brunswick: Rutgers University, 1974). Liesenbein refines Boom's Map by making the distinction between walls probably established by farmers much later and those which may have been set up in 1777.

Linn, J. B., and W. H. Egle, "Pennsylvania in the War of the Revolution, Battalions and Line," in Pennsylvania Archives, Second Series. (Harrisburg PA: L.S. Hart, State Printer, 1880).

Lossing, Benson J., *Pictorial Field Book of the Revolution*. (New York: Harper & Brothers, 1851).

Ludn, Leonard H., *Cockpit of the Revolution, The War for Independence in New Jersey*, (Princeton: Ludn Press, 2007).

Lundin, Leonard, *Cockpit of the Revolution*. (Princeton: Princeton University Press, 1940).

Lydenberg, Henry M., (ed.) *The Diaries of Archibald Robertson, Lieutenant General, Royal Engineers*. (New York, New York Public Library, 1930).

Maas, Edward J., *North of the Rariton Lotts: A History of the Martinsville Area*. (Neshanic, NJ, Martinsville Historical Committee, Somerset County Historical Society, Neshanic Printing Company, 1998).

Mackesy, Piers, *The War for America: 1775-1783*. (Lincoln: University of Nebraska Press, 1965).

Martin, Joseph Plumb, *Private Yankee Doodle*. (Boston: George F. Sheer, 1962).

Mattern, David B., *Benjamin Lincoln and the American Revolution* (Columbia: University of South Carolina Press, 1992).

Mellick, A.D., Jr., *The Story of an Old Farm or Life in New Jersey in the Eighteenth Century*. (Somerville, NJ, The Unionist-Gazette, 1889).

Menzies, Elizabeth G. C., *Millstone Valley*. (New Brunswick: Rutgers University Press, 1968).

Messler, Abraham, *Centennial History of Somerset County*. (Somerville, NJ, C. M. Jameson, 1878).

Miller, G. J. (ed.) *First Things in Old Somerset*. (Somerville, NJ: The Somerville Publishing Co., 1899).

Moran, Donald N. *New Jersey and the 1777 Forage War*. (www.revolutionarywararchives.org/newjerseyforage.html) (accessed October 9, 2017).

Muenchhausen, Friedrich, *At General Howe's Side, 1776-1778: The Diary of General William Howe's Aide-de-Camp*. (Monmouth Beach, NJ: Philip Freneau Press, 1974).

Munn, David C., *Battles and Skirmishes of the American Revolution in New Jersey*. (Trenton: Dept. of Environmental Protection, Bureau of Geology and Topography, State of New Jersey, 1976).

Bibliography 359

National Park Service, *Crossroads of the American Revolution in New Jersey*. (Philadelphia: U. S. Department of the Interior, National Park Service, Philadelphia Support Office, 2002).

New Jersey Revolutionary War Damage Claims, *Claims Against the British, Essex & Middlesex Counties*. Reel 2. (Trenton, NJ: New Jersey State Archives). There were a total of 115 individuals who filed claims for damages in Westfield during the entire war, with 92 claiming damages for June 26-27, 1777.

New Jersey, *Report of Monitoring and Recording of Crossings of Historic Walls by the Middle Brook Collector System and Supplemental Testing Adjacent to Brookside Drive for the Middle Brook Trunk Sewer*. (Trenton, NJ: New Jersey Historic Preservation Office (NJDEP), 1983).

Nixon, Richard F., *The Press in Revolutionary New Jersey, New Jersey*. (Trenton: New Jersey Historical Commission Revolutionary Experience, 1970).

Peale, Charles Willson, *"The Selected Papers of Charles Willson Peale & His Family:Volume 1" Peale: Artist in Revolutionary America, 1735-1791*. (New Haven: Yale University Press, 1983).

Peckham, Howard H., *The War for Independence*. (Chicago: University of Chicago Press, 1958).

——— *The Toll of Independence, Engagements and Battle Casualties of the American Revolution*. (Chicago: Chicago University Press, 1974).

Peterson, Harold L., *The Book of the Continental Soldier*. (Harrisburg, PA. Stackpole Books, 1968).

Petition Filed by Residents of the Vicinity of the First Middlebrook Encampment 1778. Revolutionary War Damage Claims. On file. (Trenton: New Jersey State Archives (NJDS), New Jersey).

Pluckemin, *Field Manual*, 1984, Excavating and Recording for the Pluckemin Archeological Project. Unpublished MS. (Trenton: New Jersey Archives, 1984).

Prince, Carl E., *Middlebrook, The American Eagles Nest*. (Somerville, NJ: Somerset Messenger Gazette, 1958).

Puls, Mark, *Henry Knox: Visionary General of the American Revolution*. (New York: Palgrave Macmillan, St. Martin's Press, LLC., 2008).

Raritan Landing, *The Archaeology of a Buried Port*. (New Brunswick: Rutgers Archeological Survey Office, Rutgers University, 1982).

Research and Archaeological Management, Inc., *Somerset County Cultural Resource Survey*. On file. (Trenton, NJ: New Jersey Historic Preservation Office (NJDEP), 1989).

Rice, H.C., Jr. (trans.), *Travels in North America in the Years 1780, 1781 and*

1782 by the Marquis De Chastellux. (Chapel Hill: The University of North Carolina Press, 1963).

Ricord, Frederick W., *History of Union County New Jersey 1819-1897.* (Newark, NJ : East Jersey History Co. 1897).

Risch, Erna, *Quartermaster Support of the Army: A History of the Corps, 1775-1939.* (Washington: U.S. Army, 1989).

Rodney, Thomas, *Diary of Captain Thomas Rodney, 1776–1777* (Whitefish, MT: Kessinger Publishing, LLC, 2010).

Rutsch, E. S., and R. L. Porter, 1981-82 *Draft Report, Research Materials and Field Documentation from an Archaeological Survey for the Elizabethtown Water Company Water.* Unpublished manuscript, materials on file. (Trenton, NJ: New Jersey Historic Preservation Office (NJDEP).

Sanchez-Saavedra, E. M. *Private Yankee Doodle: A Narrative of Some of the Adventures, Dangers and Sufferings of Revolutionary Soldier.* (New York: The New York Times, 1968).

Schrabisch, Max, *Excavations at Knox's Artillery Camp, 1916-1917 Pluckemin.* (Bernardsville, NJ: *Bernardsville News*).

Schleicher, William and Winter, Susan, *Somerset County, Crossroads of the American Revolution.* (Charlestown, SC: Arcadia Publishing, 1999).

Scrhrabisch, Max, "General Knox's Artillery Camp Pluckemin." (*Somerset County Historical Society Quarterly* 6 (3), 1917).

Seidel, J. L. "Archaeological Research at the 1778-79 Winter Cantonment of the Continental Army, Pluckemin, New Jersey." *Northeast Historical Archaeology,* 1983).

Seidel, John L., "The Archaeology of the American Revolution: A Reappraisal and Case Study of the Continental Artillery Cantonment of 1778-1779, Pluckemin, New Jersey.: (University of Pennsylvania, Ph.D. Dissertation,1987).

Seidel, ———— "Archaeological Research at the 1778-79 Winter Cantonment of the Continental Artillery, Pluckemin, New Jersey," *Northeast Historical Archaeology*: Vol. 12, Article 4. 1983.

Sekel, Clifford, "The Continental Artillery in Winter at Pluckemin, N. J. December 1778 to June 1779, (Staten Island: Wagner College, Master of Arts Thesis, 1972).

Sivilich, D. M., "Analyzing Musket Balls to Interpret a Revolutionary War Site." *Historical Archaeology: Journal of the Society for Historical Archaeology* Vol. 30, 1999).

Snell, J. P. (comp.) *History of Hunterdon and Somerset Counties.* (Philadelphia: Everts & Peck, 1881).

Spears, John P., *Anthony Wayne* (New York: D. Appleton & Co., 1903).
Stedman, Charles, *The History of the Origin, Progress, and Termination of the American War.* (London: printed for author by J. Murray, 1794).
Stewart, C. W., *The Life of Brigadier General William Woodford.* (Richmond, VA, Whittet Shepperson, 1973).
Stiller, C. J., *Major-General Anthony Wayne.* (Philadelphia: J.B. Lippincott Company, 1893).
Stillman, George B., "Battle of the Short Hills. Reenactment preparations June 29 and 30th, 2002," *The Brigade Courier, Brigade of the American Revolution*, Volume 18:3.
Swan, Harry Kels, *The Military Significance of Middlebrook in 1777.* Typed Manuscript. (Bridgewater, NJ: Heritage Trail Association, Van Horn House, 1977).
Stryker-Rodda, Kenneth. *Revolutionary Census of New Jersey.* (Cottonport, Louisiana, Polyanthos, 1972).
Thacher, J., *A Military Journal During the American Revolutionary War from 1775 to 1783.* 2nd ed., (Boston, MA, ,Cottons & Barnard, 1827).
Thayer, Theodore, *Nathanael Greene, Strategist of the Revolution.* (New York: Twayne Publishers, 1960).
Troiani, D. *Military Buttons of the American Revolution.* (Gettysburg: Pennsylvania.Thomas Publications, 2001).
Valis, Glenn, "The Battle of Millstone," *New Jersey in the Revolution*, http://www.doublegv.com/ggv/battles/millstone.1 (accessed October 9, 2017).
Veit, Richard Francis, *Digging New Jersey's Past: Historical Archaeology in the Garden State.* (Piscataway: Rutgers University Press. 2002).
Vermeule, Cornelius, *Revolutionary Campground at Plainfield, N.J.* (Address before Continental Chapter, Daughters of the American Revolution, January 9, 1923).
Voorhees, O. M. "Notes on Copper Mining in Somerset". *Somerset County Historical Quarterly* IV:189-194, 1915).
Ward, Harry M., *General William Maxwell and the New Jersey Continentals.* (Santa Barbara, CA: Praeger, 1997).
Washington Camp Ground Association, *George Washington Bicentennial Celebration.* (Bound Brook: Chronicle Printery, 1932).
Wayne, Anthony, Papers. (Philadelphia: Historical Society of Pennsylvania, Philadelphia).
Wickersty, Jason R., "A Shocking Havoc: The Plundering of Westfield, New Jersey, June 26, 1777," *Journal of the American Revolution*, July 2015.

Wildes, Henry Emerson, *Anthony Wayne: Troubleshooter of the American Revolution* (New York: Harcourt, Brace & Company, 1941).

Maps and Cartography

Andre, John, *Parts of the Modern Counties of Middlesex and Somerset*. Untitled manuscript map possibly authored by John Andre in 1779. (Ann Arbor: The William L. Clements Library, University of Michigan).

Beers, F. W., *1873 Atlas of Somerset County, New Jersey*. (New York: Beers, Comstock, and Cline, 1873).

Clark, J., Jr., *A Map of the Raritan River and Adjacent Country with a Plan of the Roads*. Manuscript map on file (Princeton: Special Collections, Princeton University Library, 1777). (A copy of the original map was made in 1891 and reproduced in 1931. Also see *Images of America: Somerset County, Crossroads of the American Revolution*, W. A. Schleicher, ed., and S. J. Winter. Arcadia Publishing, Charleston, South Carolina).

DeWitt, S.1780a *Contraction in the Jerseys*. Manuscript map, Erskine-DeWitt Series No. 106A, on file. (New York: New-York Historical Society). Also 1780b *Contraction in the Jerseys*. Manuscript map Erskine-DeWitt Series No. 106B, on file.

Erskine, Robert and De Witt, Samuel. *Composite of Maps of the Short Hills and Ash Swamp*. Circa 1779. Collections of the New-York Historical Society. 1777-81 Manuscript maps No. 74A, 74B, 78, on file. Can also be found at Washington's Headquarters, Morristown National Historic Park, Morristown, New Jersey and at the Alexander Library, Special Collections, Rutgers University, New Brunswick, New Jersey. The Erskine-DeWitt map series is itemized in *American Maps and Map Makers of the Revolution*, by Peter J. Guthorn (Monmouth Beach, NJ: Philip Freneau Press, 1966). Also see *British Maps of the American Revolution*. (Monmouth Beach, NJ: Philip Freneau Press, 1966).

Hopkins, G.M., and W. Kitchell. *Map of New Jersey*. (Phildelphia: H. G. Bond, 1860).

"Washington's Position at Morristown, 1780. Headquarters at Ford's House." Manuscript Map. (Ann Arbor: William L. Clements Library, University of Michigan). Also see copy at Washington's Headquarters, Morristown National Historical Park, Morristown [reproduced in Lundin, L. *Cockpit of the Revolution*, Princeton University Press, Princeton, New Jersey, 1940).

Morgan, B., *Plan of Somerset County in the Province of New Jersey*. Manuscript

map on file. (Ann Arbor: William L. Clements Library, University of Michigan, 1766).
Otley, J. W., and J. Keily, 1850 *Map of Somerset County, New Jersey: Entirely from Original Surveys*. (Camden, NJ:, 1850).
Scull, W. *Road from Quibbletown to Amboy*. Manuscript map on file. (New York, New-York Historical Society). [Erskine-DeWitt Series No. 55, 1779].
Reid, J., *A Mapp of Rariton River*...Manuscript 1686 on file. (Newark: New Jersey Historical Society).

Index

A

Abeel, James, 187
Abercromby, Robert, 11, 14, 18
Ackly, Bezeliel, 288
Adams, Abigail, 65
Adams, John, 65, 264, 287
Adams, Samuel, 298
Albany, NY, 153
Alexander, William, Lord Stirling, 82, 89, 268, 284
Allen, Ethan, 262
Alliance with France (Treaty of Alliance 1778), 134, 161, 165
Amboy Bay, 235
American Eagles Nest, C. Prince, 258
American Legion Post 63, Bound Brook, 238
Andre, John, 57, 83, 99
Angelakos, Peter, 258
archeology, 256-257, 268, 309-310, 312, 317
Armand-Tuffin, Charles, Marquis de la Rouërie, 84, 87, 94
armorers, 272-274, 278, 302, 305
Arnold, Benedict, 160, 262
Arnold, Jacob, 5
Arthur Kill, 75, 78, 80, 101, 233
artificers, 274, 277-279, 295, 302, 305 shop, 177
artillery, 18, 20-22, 29-30, 39, 42-44, 46, 53-54, 58-60, 72, 82-84, 86-87, 90, 94, 103, 135-136, 140, 162, 165, 177, 239, 250, 262-263, 265, 270-274, 276-278, 280-281, 285-286, 299-300, 302-303, 308, 312
Ash Swamp, 73-74, 76, 86-88, 90, 92, 96-97

B

Baldwin, Jeduthan, 165, 177, 278
Balmain, Alexander, 221
Barber, Francis, 117, 142
Basilone, John, 238
Basking Ridge, NJ, 44, 73, 284
Beatty, John, 173, 221
Bedminster Dutch Reformed Church, 291
Bedminster Township, NJ, 188, 262, 273, 291-292, 307, 310
Bedminstertown, NJ, 304
Bergen County, NJ, 208, 284
Bernardsville News, 268, 305-306
Betsy Ross Flag, 69
Bibles, distribution of, 141
Biddle, Clement, 165
Biddle, Owen, 188
blacksmiths, 278
Bloody Gap, 84, 90, 95, 104
Bloomingdale, NJ, 171
Blue Hills Post, 73
Board of War (and Ordinance), 45, 189, 192
Bogert, Guysbert, 15
Bolmer Farm, 53, 58, 208
Bolmer, Kennedy, 208
Bonhamtown, NJ, 5, 10, 31, 33

Index

Boom, A.A., 64, 210, 234, 250-254, 256
Boston Massacre, 286
Boston, MA, 188-189, 192, 263-264, 275, 286
Boston, siege of, 211
Bound Brook, 111-112, 114
Bound Brook Elizabethtown Reservoir, 254
Bound Brook, NJ, 4, 10, 12, 16, 22-24, 27, 35-44, 47-48, 55, 59, 67, 73, 90, 98, 111, 123-124, 157, 166, 177, 182-183, 189, 204, 211-212, 215, 217, 226-228, 237, 255, 258, 267, 317
    battle of, 6, 35, 37, 41, 43-47, 49-51, 62, 225-227
    hospital, 298
    slaughterhouse, 297
Bower, Ernest R., 255-256
Brandywine, battle of, 9, 148, 211
Bridgewater Township, NJ, 110, 121-122, 254, 317
Bridgewater, NJ, 5, 36, 41-42, 51, 64, 109, 115, 121, 123, 143, 164, 176-177, 215, 217, 221, 229, 318
Britain declares war on France (1778), 135, 157
British Army, 3-11, 17, 24-25, 28-30, 32-33, 35, 37-38, 40, 42-45, 50-51, 54-55, 62, 64-65, 67, 69-72, 75, 78, 80-83, 85, 87-89, 91-93, 95-104, 107, 110, 112, 118, 153-154, 162, 174, 187, 201, 207, 210-211, 214, 319
British Articles of War, 168
British *New York Gazette*, 45
Bronx, NY, 211
Brooklyn, NY, 99
Brunswick, NJ, 10, 15, 18, 29
Buffalo, NY, 159

Bullion's Tavern, 215
Bunker Hill, battle of, 286
Burgoyne, John, 54, 103, 114
    surrender at Saratoga, 161
Burlington, NJ, 22-23
Burns, Brendan, 121
Burnt Mills, 188
Butler, William, 165

C

Calco Chemical Company, 228
Cambridge, MA, 224, 264, 286
camp followers, American, 146-148, 151, 154
camp followers, British, 153-154
camp followers, slaves, 152
celebrations for the Army, 161-162
Centennial History of Somerset County, A. Messler, 59, 258
Central Jersey Regional Airport, 21, 231
Charles R. Ware (US Destroyer), 238
Charleston, SC, 184, 213
Chesapeake Bay, 78, 101-102, 184, 214, 244
Chimney Rock, 233-234, 239, 250
Chimney Rock Pass, 51, 53, 58-59, 111-112, 124, 208, 217, 317
Church of England, 98
Church, John, 183
Clark, Elizabeth, 154
Clinton, Henry, 102, 211, 214, 284, 315
Clinton, James, 224
Clothier Generals Depot, 188
Clothier's Office, 178
clothing supplies, 188-189
    shoes, 152, 188-189
cockfights, 295
Coleman, James, 172

Colfritt, Mary, 154
Colles, Christopher, 279
Commander in Chief's Guard "Washington's Life Guard", 128, 162, 224, 293
comte de Rochambeau, Jean-Baptiste Donatien de Vimeur, 122, 214
Concord, battle of, 31, 191, 286
Connecticut Farms, NJ, 27, 159, 247-248
construction, 7, 41, 47, 56, 113, 120, 124, 127-128, 131, 210, 251, 273, 275-278, 287, 294, 301-302, 307
Continental Army, 3-10, 24-27, 32-37, 44-45, 47, 49-51, 53-56, 63, 65-67, 69-70, 73-74, 76, 78, 80-82, 84, 86, 89-91, 95-96, 101-103, 107-108, 110, 112-114, 116, 118, 127, 134-135, 137, 162, 191, 206, 209-211, 216, 221, 224-226, 229, 232, 235, 242, 248-249, 252-253, 267, 287, 293, 296, 300-301, 313-314
Continental Articles of War, 170, 173
Continental Artillery, establishment of, 266
Continental Congress, 6, 10, 65, 69, 95, 141, 153, 158-159, 182, 264, 268
Continental Congress, Second, 68, 116, 125, 136-137, 171, 194, 223, 228, 287
Convention Army, 160
Conventions and Congresses of Massachusetts Bay, 45
Conway, Thomas, 84, 87, 91-92
Cornwallis, Charles, Lord, 3-4, 14, 23, 28, 30-31, 38, 43, 45, 83-96, 104, 214, 227, 242, 300
corporal punishment, 50, 136, 140-141, 167-174

corporal punishment, civilians, 154
Corps of Artificers, 163, 165, 176-177
counterfeiters, 197
court martials, 156, 163, 167-169, 174, 295
Cox, William, 167
Cresswell, Nicholis, 71
crimes commited in encampment, 167-168
Cronkite, Maria, 151
Crossroads of the American Revolution, National Heritage Area, 122
Cultural Resources Consulting Group, 254
Cumberland County, NJ, 15
currency, Continental, 54, 185-187, 191, 197, 230

D

Danbury, CT, mutiny of, 118-119, 289
Darke, William, 83, 84
Davis, T.E., 258
Dayton, Elias, 87, 93
De Mott's Tavern, 213
Declaration of Independence, 69
DeKalb, Baron Johanne, 108, 116, 118, 123, 230-231
Delaware & Hudson Canal, 235
Delaware & Raritan Canal, 47, 226
Delaware River, 25, 50, 63, 110
Dennins, Robert, 167
desertion of soldiers, 54, 60, 138, 156, 162, 167, 170, 172, 209, 288, 290, 298
Dickinson, Philemon, 8, 12-15
Dilkes, Andrew, 18
Dismal Swamp, 10, 85
Dock Watch Hollow, 58
Donavan, Timothy, 295

Dorchester Heights, battle of, 264
Doughty, John, 288
Drake, Nathanial, home of, 75, 82
Draper, George, 183
Drew University, 307
drumming out of the Army, 154, 173
drunkenness in camp, 156, 163, 167, 199, 288, 295
Duer, William, 281
Dunn, Joseph, 221
DuPortail, Louis, 231
Dutch Meeting House, 212

E
Eagle's Nest Museum, 60, 236-237, 239
East River, 99
Easton, PA, 209
Edison, NJ, 33, 70, 72-73, 76, 80-83, 85
Elizabeth, NJ, 56, 208, 215
Elizabethtown Water Company, 254
Elizabethtown, NJ, 112, 117-119
Elkton, MD, 78
Englishtown, NJ, 212
entertainment in encampment, 138-139
Eoff, Jacob, 304
Erskine, Robert, 124, 178, 259
Erskine, Sir William, 30, 57, 64
Erskine-DeWitt army map, 250, 259
Essex County, NJ, 81, 114, 208
execution of soldiers, 140-141, 168, 172, 289-290

F
Fanwood-Scotch Plains Rotary Club, 88
Federal Hill Rebellion, 171

Ferguson, Patrick, 83
Ferguson, William, 46
Finch, John, 86, 90, 94
Finderne, Bridgewater, NJ, 16, 36, 41, 143, 176, 194, 212
First Church of Raritan, 89, 233, 258, 316
Fisher, Elijah, 128, 162
Flag Resolution, 68
Flint, Royal, 180, 297
flogging as punishment, 173, 174
Folcard Sebring House, 51
Food, 8, 11, 13, 32, 46, 60, 62, 98-99, 128, 130, 181-182, 184-185, 189, 193, 199, 204-205, 276, 297, 300
Forage Wars (1777), 5, 8, 25, 82, 102-103, 111-112, 199, 201, 212
Forest, Thomas, 270
Forsythe, Robert, 165
Fort Montgomery, 139
Fort Niagara, 159
Fort Pitt, 209
Fort Ticonderoga, 262-264, 275, 287
Framingham, MA, 264
Franklin Township, NJ, 19, 41, 63, 111, 212-213
Franklin, Benjamin, 179, 225
Frazee, Elizabeth, 88
Frazee, Gershom, 87
home of, 88, 95
Frederick the Great, King of Prussian, 225, 314
Fredericksburg, NY, 269
Freehold, NJ, 102, 107
Frelinghuysen Tavern, 212
Frelinghuysen, Frederick, 73
Friends of Abraham Staats House, 226
Frothingham, Richard, 272-273

G

gallows, 168, 172
gambling amongst soldiers, 295
Gano, John, 144
Garrett-Voorhees farmhouse, 212
Gates, Horatio Lloyd, 108, 116
Gibbs, Caleb, 222
Gibson, John, 165
Girard, Conrad Alexandre, 135, 157-158
Glenn, Thomas, 16
Grant, James, 31, 39, 42, 45, 90
Great Lakes, 159
Green Brook, 10, 43, 73-74, 76, 80-81
Green Brook Rock, 233-234
Green Brook Township, NJ, 73, 232, 318
Greenbrook Pass, 27, 88, 112
Greene, Catherine Littlefield "Kitty", 151, 165, 192, 229, 281
Greene, Nathanael, 44-45, 47, 50, 57, 60, 62, 70-72, 78, 84, 123-124, 127, 133, 152, 157, 165-166, 176-177, 179, 184-186, 189-195, 197-198, 204, 221, 227, 229, 268, 299, 315
Griffin, Simon, 71
Griggstown, NJ, 21, 23
Guest, Moses, 213

H

Half Moon Battery, 39-40, 43
Hall, Josias, 123
Hamilton, Alexander, 100, 223
  as Secretary of the Treasury, 283
Hancock, John, 13, 35, 69
Hancock, William, 211
Harcourt, William, 41
Hardenbergh, Rev Jacob Rutsen, 143
Harlingen, Earnestus, 205-206

Harrison, Benjamin, 125
Hartley, Thomas, 256
Haverstraw, NY, 107, 118
Hawk's Watch (Chimney Rock), 233
Hays, William, 148
Helnit, William, 289
Henry, Patrick, 188
Herbert, John, 228
Heritage Trail Association, 229
Hessian troops, 4, 5, 8-12, 15, 17-18, 20, 22, 25, 29-30, 37-42, 44-45, 47, 50, 54, 72-73, 81-82, 86-87, 90-91, 94, 98
  Grenadier Battalion, 29-30
  Jaeger Battalion, 28
  Kessel Company, 29
  Minnigerode Grenadiers, 93
Highland Park, NJ, 70
Hill Survey (1754), 64
Hills Development Corporation, 307-309
Hills, John, 259
Hillsborough Dutch Reformed Church, 24, 143
Hillsborough, NJ, 14, 18-19
Historic Conservation and Interpretation, Inc., 254
Hodgdon, Samuel, 302-303
Honeyman, John, 23
Hooper, Robert, 186
Hopkinson, Francis, 69
horse corral, 177, 183
hospitals, 84, 145, 151, 153, 183-184
housing of soldiers, 129, 210, 221, 262, 299, 318
Houston, William Churchill, 16
How, David, 158
Howe, Sir William, 3, 26, 40, 45, 54-56, 60-65, 67-70, 74-76, 78, 80,

Index

82-85, 88, 90-92, 95, 97-98, 100, 102-103, 114, 232, 264
Hudson Highlands, 4, 107, 117-119, 152, 209
Hudson River, 103, 107-108, 117-118, 129, 211, 269
Hudson River Valley, 103, 208
Hunt, Benjamin, 295
Hunter Research, 253, 256-258, 309, 318
hutting, 119, 124, 127, 136, 163, 315

I
infant mortality, 202
Iroquois Nation, 135, 137, 159, 188, 190-191, 209, 231, 298, 315
Irvine, William, 123

J
Jacobus Vanderveer House, 221, 230, 262, 286, 291, 300, 309-310
  Friends of, 292
Jockey Hollow, encampment, 131, 169, 172, 190, 239, 247, 285, 298, 301, 306, 315, 318
John Basilone Parade, 238
Johnson, John, 304
joiners, 278
Jones, William, 182

K
Kings Ferry, NY, 118
Knox, George, 286
Knox, Henry, 123, 160, 164-165, 181, 221, 230, 262-266, 268-270, 274-276, 279-284, 286-292, 294-295, 297-303, 312, 318
  as Secretary of War, 283, 300
  death of, 300

Knox, Julia, 291
Knox, Lucy, 151, 165, 281, 286-287, 291, 300

L
Lafayette, Marquis de, 133, 315
Laing Farm, 94
Lake Champlain, 263
Lake George, 263
Lamb, John, 268, 270, 288, 295-296
Lambert, James, 88
Lambert's Mills, 88
Lamington River, 178
Lang, Steven, 121
Laurens, Henry, 125, 281
Lee, Henry, 212
Lehigh County Historical Society, 132
Lenape Tribe, 307
Leslie, Alexander, 29
Leslie, William, 100
Lexington, battle of, 31, 191, 286
Liberty Corner, NJ, 207, 215
Library of Congress, 259
Liesenbein, William, 252-254, 256
Lillie, John, 307
Lincoln, Benjamin, 36-37, 42-46, 57-58, 62, 84, 108, 116, 227
Lincoln's Gap, 58
liquor, 23, 28, 154, 156, 161, 163, 168, 205, 216, 288, 295
  sales in camp, 163-164
Livingston, Betsy, 165
Livingston, Susannah French, 165
Livingston, William, 100, 159, 165, 200, 205, 210-212, 221, 228, 284
London Annual Register, 65
Long Hill Township, NJ, 27
Long Island, NY, 27
Lossing, Benson, 14, 123, 234-237,

240, 252, 257
Lotts, Cornelia, 165
Lowantica Valley, encampment, 5
Ludlow, Henry, 292
Ludlow, Jacob, 98

M

Madden, John, 64
Maddox, Mary, 222
Maitland, John, 42-43
Manley, Adrian, 121
Manville, NJ, 8, 12, 19-20, 22, 111, 123, 215, 316
  Public Library, 12
Mariner, William, 213
Martin, Joseph Plumb, 186, 313
Martin's Woods, 85
Martinsville VFW Post 1388, 238
Martinsville, NJ, 27, 51, 54-55, 58, 60, 89, 110-111, 124, 208, 238, 317
encampment, 250
Matonaking, NJ, 90
Matthews, David, 45
Mawhood, Charles, 31-32
Maximus, Quintus "Fabius", 65
Maxwell, William, 32-33, 57, 86-87, 91-93, 209, 248
McBride, Don, 240
McBride's Rock, 240
McCauley, Mary Ludwig Hays, 148
McDougall, Alexander, 108, 116, 193, 276, 297
McHenry, James, 223
Meheim, John, 189, 200
Mellick, Andrew D., Jr., 227
Messler, Abraham, 59, 89, 233, 258, 316
Metuchen Meeting House, 92
Metuchen, NJ, 10, 30, 33, 72, 85, 90,
92, 104
Middle Brook, 51, 53, 55, 57, 64, 111, 119, 121, 139, 202, 207-208, 228, 249, 256, 318
Middle Brook Creek, 121
Middlebrook Campground, 249-250, 253, 257-258, 268, 291
  Association of, 237
Middlebrook Rock, 233, 234
Middlebrook, NJ, 5, 11, 27, 55-56, 62-63, 66, 68, 75, 84, 95-96, 119, 124, 217, 232-234, 267
encampment, 3, 7, 25, 27, 44, 47, 49-51, 53, 55-57, 59-61, 63-65, 67-68, 70, 72, 75, 78, 80-81, 90, 100-102, 107-108, 110-112, 114, 116-120, 123-125, 127-128, 130, 133-134, 136, 138, 143, 145, 156, 160, 167-168, 176, 178-179, 181, 183-184, 188, 190, 196, 198-199, 207-209, 214-215, 217, 220, 222, 225-226, 228-230, 232, 234-235, 237-238, 249, 254, 257-258, 267, 269, 273, 280, 285, 291, 298, 313-317
Middlebush, NJ, 67
Middletown, NJ, encampment, 169
military units
  1st Connecticut Militia, 10
  1st Maryland Regiment, 167
  1st Massachusetts Regiment, 140
  1st New Jersey Regiment, 73, 93-94
  1st New York Regiment, 151
  1st Pennsylvania Brigade, 256
  1st Pennsylvania Division, 57
  1st Virginia Regiment, 167
  28th Battalion, 29
  2nd New Jersey Regiment, 94, 100
  2nd Pennsylvania Brigade, 57
  2nd Pennsylvania Regiment, 150
  3rd Continental Artillery, 298

Index

3rd Maryland Regiment, 152
3rd New Jersey Regiment, 93-94
3rd Pennsylvania Brigade, 87, 91-92
4th Continental Artillery, 37
4th New Jersey Regiment, 94
4th Pennsylvania Regiment, 148
5th New York Regiment, 139
5th Virginia Regiment, 30
6th Pennsylvania Regiment, 148
8th Pennsylvania Regiment, 36
8th Virginia Continental Line, 10
Artillery Brigade, 262
Continental Dragoons, 212
Corps of Artillery, 123
East Jersey Artillery Co., 93-94
First Middlesex Regiment, 73
First Somerset Regiment, 73
Green Mountain Boys, 262
Light Infantry Corps, 167
Maryland Brigade, 123
Maryland Division, 124, 230
Maryland Regiment, 133, 204
Morgan's Rifle Corps, 83, 92
New Jersey Brigade, 57, 87, 91-95, 117, 171, 224, 316
New Jersey Militia, 6-7, 28-29, 32, 54, 56, 72-73, 96, 112, 139, 159, 189, 201, 248, 319
New York Brigade, 131, 139, 144, 217
Ottendorf's Corps, 85, 94
Pennsylvania Brigade, 55, 57, 91, 93, 123, 239
Rhode Island Militia, 191
Somerset County Militia, 8, 16
West Jersey Artillery Company, 93
military units (British)
1st British Grenadiers, 86, 92
2nd Light Dragoons, 92
37th Foot, 11

42nd Highlanders, 30, 32
British Light Dragoons, 92-94, 187
Queen's Rangers, 86, 143, 211-213, 228, 316
Royal Highlanders, 29
Millington, NJ, 27, 207
Millstone Creek, 22-23
Millstone Forge Association, 24
Millstone River, 4, 5, 8, 10-12, 14-23, 26, 28, 36, 63-64, 80, 139, 202
Millstone River Valley, 4, 27
Millstone, NJ, 9-10, 12, 14, 16, 23, 33, 35, 63, 67-68, 72, 111, 123, 143, 179, 183, 203, 205, 212, 215, 217, 221
battle of, 8, 24, 26
Miralles, Don Juan, 157
Monmouth County, NJ, 187, 208
Monmouth Court House, NJ, 27, 195, 314
Monmouth University, 308-309
Monmouth, battle of, 102, 107, 116, 148
More, John, 295
Morgan, Daniel, 118
Morris County, NJ, 56, 112
Morristown Historical Park, 259
Morristown National Historic Park, 307, 310
Morristown, NJ, 4-5, 9, 25, 27, 35, 37, 42, 51, 55, 62, 97, 111-112, 179, 188, 190, 195-196, 201, 203, 205, 210, 215, 232, 300
encampment, 4-5, 9, 11, 33, 36-37, 44, 49-50, 65, 110, 257, 303
mutiny, 170
Quartermaster store, 195-197
Mott, John, 289
Mount Prospect Camp, 58
Mount Vernon, 125, 300

Mountainside, NJ, 98
Muhlenberg, John Peter Gabriel, 57, 121, 165

N
National Historic Register, 230
National List of Historic Places, 292
National Museum of the American Revolution, 55, 238
National Park Service, 122, 217
National Register of Historic Places, 41, 224, 226, 229, 284, 317
Navesink Mountains, 235
Neshanic, NJ, 28
New Bridge, NJ, 22-23
New Brunswick, NJ, 4-5, 8-9, 11-12, 14, 16, 24-26, 28, 30-32, 35-36, 38, 40, 44-46, 53, 55-56, 59, 62-64, 67, 69-72, 78, 80, 82, 97, 111-112, 162, 183, 201, 212-213, 220, 232, 235, 239, 259, 316
  British encampment, 6, 110
  hospital, 298
New Jersey Campaign, 97
New Jersey Historic Preservation Office, Trenton, NJ, 255
New Jersey Legislature, 68, 205
New Market, NJ, 10
New Windsor, encampment, 50, 110, 131, 257, 296, 301, 318
New York City, 4-5, 50, 99, 102-103, 107-110, 112, 118, 120, 189, 207, 211, 214, 220, 243, 246, 264
New York Harbor, 103
New York Historical Society, 259
Newark, NJ, 107, 178, 190, 215
Newton, NJ, 254
Nielson, John, 13
North of the Rariton Lotts-A History of the Martinsville, New Jersey Area, 250, 254
North Plainfield, NJ, 27, 65, 73, 80, 84, 89
nurses, army, 146, 152-153

O
O'Brien, Anne, 306
Oak Ridge Park, 92
Oak Tree, 73, 83-85, 93-94
Old Dutch Parsonage, 224
  Association of, 224
Old Millstone Forge, 23
Old Stone Arch Bridge, 41
Oliver/McCain/Campbell farmstead, 254, 257
Ontario, Canada, 159
Order of the Bath, 60
Otley, J.W., 64

P
Paramus, NJ, 184
Parsippany, NJ, 159
Parsons, Samual Holden, 84
Passaic County, NJ, 302
Patullo, Benjamin, 237
Patullo, Herbert M., 60, 236-240
  obituary, 237
Patullo, Marianna, 237
Peale, Charles Willson, 54, 76, 125
Peebles, John, 32
penal colony (Australia), 175
Pennsylvania Council, 160, 197
Pennsylvania Evening Post, 46, 97
Pennsylvania Packet or the General Advertiser (Philadelphia), 274, 281, 310
Perth Amboy, NJ, 5, 9, 24, 32, 64, 71-72, 78, 80, 82-83, 85, 101

Index 373

Pettit, Charles, 128
Philadelphia & Reading Railroad, 22
Philadelphia Campaign, 96, 102
Philadelphia, PA, 3-4, 6, 41, 44, 47, 50-51, 62, 64, 69-70, 78, 89, 96-97, 101-102, 107, 110, 114, 120, 125, 154, 160, 166, 179, 184, 192, 194, 203, 223, 232, 239
Pictorial Field-Book of the Revolution, B. Lossing, 234
Pierce, John, 186
piety amongst soldiers, 141-145, 163
Pillar of Fire, 47
Piscataway, NJ, 10, 27-28, 33, 41-42, 67, 72, 76, 78
Piscatawaytown, NJ, 5
Pitcher, Molly, 148, 296
Plainfield Country Club, 73, 81, 85-86, 94, 103
Plainfield, NJ, 75, 80-82, 85
Pluckemin Archaeological Project, 307
Pluckemin Artillery Camp, 129, 148, 164, 249, 258, 262, 267-268, 270, 272-280, 284-285, 287-288, 291-297, 299-310, 312, 318
  Academy, 267, 274, 275, 279-281, 283-284, 291, 294, 301, 307, 311
  hospital, 298-299, 303
  mutiny of, 290-291
Pluckemin, NJ, 5, 51, 59, 111, 121, 123, 129, 143, 147-148, 165, 168, 177-178, 181, 183-184, 188, 196, 199, 207-208, 221, 230, 233-235, 249, 262, 266, 269, 273, 284, 291, 295, 303, 305, 318
Pompton, encampment, 50, 110, 144, 300, 302
  mutiny of, 170-171
Pope's Day, 144

Porter, Richard, 254
Post Office, 178
Presbyterian Church at Bound Brook, 143
Presbyterian Church Burial Grounds, 88
Prescott, Richard, 29
Pride, James, 139
Pride, Mahitable, 139
Prince, Carl E., 258
Princeton University, 16
Princeton, NJ, 4, 8, 11, 14, 112
  battle of, 4, 11, 18, 23, 25, 31, 33-34, 50, 55, 110-111
Proctor, Thomas, 42-46, 148, 296
Putnam, Israel, 108, 116

Q
Quartermasters Stores, 179
Quebec City, Canada, 50, 110
Queens Bridge, 36, 40, 47
Quibbletown Gap, 27, 49, 74, 84, 285
Quibbletown, NJ, 10, 27-31, 57, 73, 76, 78, 84, 91, 93, 212

R
Rahway River, 32, 99
Rahway, NJ, 32, 80, 82, 101
Ramapo Hills, 4
Raritan Bay, 70, 112, 235
Raritan Landing, 28-30, 42, 44, 71
  Bridge, 36
Raritan River, 10-11, 18-20, 22-23, 28, 36-38, 40, 43, 46, 62-64, 69, 71, 98, 103, 107, 111, 114, 120-121, 123, 139, 178, 202-203, 207, 209, 211-212, 221-222, 228-230, 233, 235, 273, 291, 316
Raritan River Valley, 26-27, 33, 54, 63, 112-113, 119-120, 183-184, 190, 199-201,

207, 211, 214, 216-217, 224, 229, 249, 274, 292, 316
Raritan, NJ, 111, 128, 173, 180, 183, 224, 238, 278
ration of food, definition, 180
Reed, Joseph, 69, 75, 129, 197
Rees, John U., 146
Reformed Dutch Church of Raritan, 143, 206
Regulations for the Order and Discipline of the Troops of the United States, "Blue Book", F. von Steuben, 134, 293
Revolutionary Memorial Society, 224
Risch, Erna, 187
Robertson, Archibald, 14, 33
Rockland County, NY, 118
Rodney, Thomas, 11
Round Top, 59
Rowe, Michael, 288
Royal Artillery Academy, 301
Royal Gazette, 220
Royce Brook, 20
Rush, Benjamin, 58
Rutgers University, 258-259, 269, 307
    Alexander Library, 259

S

Sacred Heart Cemetery, 19
Salem, NJ, 211
Samptown, NJ, 10, 27-31, 65, 76, 78, 84-85, 90
Sandy Hook, NJ, 102, 107, 162, 235
Saratoga, battle of, 92, 103, 160-161, 269
Sayreville, NJ, 212
Scammell, Alexander, 117, 142, 165
Schenk, Hendrik, 21
Schley, Grant Baker, 292-303, 305

Schrabisch, Max, 258, 268, 304-306
Scotch Plains, NJ, 49, 70, 73, 76, 80-82, 84, 86-88, 90, 95, 97, 100, 104, 107
Scott, Charles, 30-31, 57, 121
Scott, John, 10
Scull, William, 57, 64, 235
Scully, William, 167
Seidel, John, 258, 269, 307-310, 312
Sekel, Clifford, 258, 268-269, 273, 278, 306-307, 310
Selig, Robert, 122, 217
Selody Sod Farm, 20
Shandor, Arthur, 12, 19, 22
Shandor, Fred, 121
Shippen, Margaret "Peggy", 160
shoemakers, 190, 278
Short Hills, 24, 76, 80, 82, 84-85, 92-93, 98
Short Hills Battleground Historic District, 93
Short Hills Tavern, 86, 94
Short Hills, NJ, 73, 81
    battle of, 6, 49, 77-78, 81, 87, 90-91, 95-97, 99-101, 103, 114
Shreve, Israel, 100
Simcoe, John Graves, 23, 143, 211-214, 228, 316
Singer Company, 230
slaughter house, 62, 182
smallpox, 5, 26, 33
Smallwood, William, 123, 133, 164
Smith, Hugh, 168, 178
Society of Colonial Wars, 258
Somerset Community College, 307
Somerset County Historic Society, 230
Somerset County Historical Quarterly, T. Davis, 207, 258, 268

Somerset County Park Commission, 256
Somerset County Park System, 115
Somerset County, NJ, 27, 33, 44, 47, 51, 56, 64, 111-112, 119, 184, 188, 199, 201, 208, 216, 220, 226, 228, 239, 253-255, 284, 292, 317
Somerset Court House, NJ, 9, 11, 14, 23, 63, 177, 179, 183, 212
Somerset Hotel, 179
Somerset Land Map (1850), 64
Somerset Patriot Baseball Park, 111
Somerville, NJ, 111, 125, 128, 143, 151, 157, 177-178, 220, 222, 238
Sons of Liberty, 286
South Amboy, NJ, 212
South Bound Brook Historic Preservation Advisory Commission, 226
South Bound Brook, NJ, 36, 111, 221, 224, 226, 231
South Branch, 28
South Plainfield, NJ, 10, 27, 29, 63, 65, 73, 76, 78, 85, 104, 215
Spanktown, battle of, 31-32
Sparkston, NJ, 90
Spaulding, Simon, 42
Spencer, Joseph, 9
Spencer, Oliver, 67
Spring Valley Road, 5
Springfield, NJ, 27-28, 35, 55, 70, 73, 208, 248
St. Clair, Arthur, 123, 159, 197, 231
St. Paul's Lutheran Church, 143
Staats House, 157, 221, 224-226, 231
Staats, Abraham, 221, 224-226
Stage House Tavern, 213
Stark, John, 153
Staten Island, NY, 64, 70, 74-76, 78, 80, 89, 91, 95, 101, 109, 117, 212, 233, 235
Stavola Industries Quarry, 239
Stedman, Charles, 9
Steele's Gap, 121, 124, 285
Steeles Tavern, 121, 124, 163, 217
Stempien, David, 121
Stephen, Adam, 84
Stevens, Ebeneser, 270
Stevens, Edward, 57, 282
Stewart, Bruce W., 307
Stillman, George W., Sr., 90
Stillwell, George, 87
Stirling, William Alexander, Lord, 57, 70-78, 80-82, 84-88, 90-96, 100, 103, 108, 114, 116-117, 127, 129, 141-142, 160, 187, 207, 221, 228, 242
Stone Brook, NJ, 11
Stony Brook Pass, 27, 74, 80, 84, 89
Stony Point, battle of, 209, 294
Story of an Old Farm, A. Mellick, 227
Strawberry Hill, 83-84, 92
Stryker, John, 8-9
Stuart, Sir Charles, 33, 90
Sugar House Prison, 99
Sullivan, John, 84, 159, 209, 224, 231, 315
Sullivan, Thomas, 17, 29
Sullivan-Clinton Expedition, 159, 188, 190, 298
supply shortages, 99, 109, 114, 134, 148, 176, 180, 182, 187, 189, 191-195, 197, 203, 209, 220, 230, 303, 313, 315
Sussex County, NJ, 56
Sutphen, Samuel, 15-16

T
tailors, 177, 278
Tarleton, Banastre, 83

Tarrytown, NY, 269
Temporra, Victor, 240
Ten Eyck House, 178
Ten Mile Run, 8
Thacher, James, 66, 129, 131, 163-165, 171-172, 174, 197, 280
Thomaston, ME, 300
Thompson, Thomas, 289
Tilghmam, Tench, 223
Tin Can Sailors - The National Association of Destroyer Veterans, 238
Trenton, NJ, 4, 25, 112, 166, 184, 190, 195-196
battle of, 4, 18, 25, 33-34, 50, 110
magazine, 195
Tunison, Cornelius, 178
Tunison's Tavern, 178
Turnbull, Charles, 46

U
United States Military Academy, West Point, 267, 283, 312
University of Michigan, 259
University of Pennsylvania, 307

V
Valley Forge, encampment, 49-50, 107, 110, 120, 127-128, 131, 133-135, 141, 153, 161, 179, 190, 192, 194-195, 203, 225, 239, 257, 293, 301, 306, 315, 318
Van Doren House, 23
Van Horn, Harriet, 165
Van Horne House, 42-43, 177, 221, 226-228
Van Horne, Philip, 42, 212, 226-227
Van Middleswoth's Island, 231
Van Nest House, 221
Van Nest, Abraham, 21, 22, 24, 204

Van Nest, Catherine, 24
Van Nest, John, 196
Van Nest's Mill, 8, 11-13, 15, 17, 19, 21-24, 26, 28, 123
battle of, 11, 19, 22
van Ottendorf, Nicholas, 84
Van Veghten Bridge, 16-18, 36, 41, 43, 211-212, 215, 316
Van Veghten House, 21, 166, 176, 179, 189, 194, 221, 229-230
Van Veghten, Derrick, 229-230
Van Veghten, Michael, 229
Vanderveer, Jacobus, 221, 291, 304-305
Vaucher, Bob, 121
Vaughan, John, 83, 85, 92, 95
Veit, Richard, 308
Vermeule Camp, 73-74, 76, 81
Vermeule, Cornelius, 65
home of, 74
plantation of, 65, 73
Virginia House of Burgesses, 125
von Donop, Carl, 29, 39-40, 42, 45
von Ewald, Johann, 10, 22-23, 30, 38-40, 43-44, 47, 95
von Knyphausen, Wilhelm, 247
von Minnigerode, Colonel, 86
von Muenchausen, Friedrich, 63, 69, 90, 102
von Steuben, Friedrich Wilhelm, Baron, 128, 134, 157, 209, 221, 224-226, 293-294, 313
Voorheis, Cornelius, 200

W
Wadsworth, Jeremiah, 180, 229
waggoner, 295
Wallace House, 125, 128, 143, 151, 157, 159, 178, 221, 222-224, 231

Wallace, John, 222
Wallace, Mary, 220, 222
Ward, Artemas, 286
Warner, Nathaniel, 31
Warren Township, NJ, 27, 111, 207, 215, 285
Warren, NJ, 51, 80, 89-90, 95
Washington Campground Association, 68, 114, 237, 240, 317
Washington Rock, 60, 73-74, 76, 81, 232-234, 237, 241, 244-247, 252
Washington Rock State Park, 318
Washington Valley, 27, 51, 55, 58-59, 63, 74, 80, 89, 111, 113, 119-120, 122, 124, 184, 199-201, 207-208, 214, 216, 239, 249, 251, 256, 317
Washington Valley Park, 254, 317-318
Washington, George, 3-5, 7, 9-11, 13-14, 23, 25-27, 33, 35, 37, 44-47, 49-51, 54-55, 57, 59-60, 62-70, 72, 75, 78, 80-84, 88, 90-1, 95, 97, 100-103, 109, 111, 114, 116-119, 124-125, 127-128, 133-134, 136, 138, 141, 146-147, 150, 153-154, 156-159, 162-163, 165-166, 168, 172, 175, 181, 183, 187, 192-193, 195, 200, 203, 205-207, 209-210, 214-215, 220-223, 226, 228-229, 232-233, 235, 238, 241-243, 262, 268, 280, 283-84, 286, 298, 313, 315
  appearance of, 66
  as President, 283, 300
  nicknamed "American Fabius", 65-67, 69
Washington, John Augustine, 55, 63
Washington, Martha, 125, 146, 151, 165, 192, 222-223, 281, 300
Washington's Rock, 232, 234, 242, 244
Washington-Rochambeau
    Revolutionary Route National Historic Trail Project, 122, 217
Watchung Mountains, 4, 7, 10, 24, 27, 31, 35, 37, 42-44, 49, 51, 53-55, 57, 64-65, 73-75, 77, 80, 82, 89-90, 95-96, 103-104, 109, 111-112, 119-120, 195, 202, 207, 232, 241, 247-249, 253-254, 256, 262, 273, 284, 317-318
Watchung, NJ, 27, 80, 84, 89-90, 95, 104
Waxhaws, battle of, 83
Wayne, Anthony, 43, 53, 55, 57-60, 63, 81, 108, 118, 123, 129, 196, 204-205, 209, 221, 231, 235, 239, 249, 250, 253-256, 258, 268, 316, 318
Wayne's B, 57
Wayne's Gap, 55, 58, 285
Webster, John, 73
Weedon, George, 57
Weiss, Jacob, 132, 200
Wemple Homestead, 121-122
West Point, NY, 118-19, 210, 269, 300
Westchester County, NY, 108, 269
Westfield Meeting House, 98
Westfield, NJ, 82, 86, 88-90, 95, 97-101, 215
Weston, NJ, 8, 14
wheelwrights, 177, 278
White Plains, NY, 27, 108, 116
  encampment, 116, 118
Whitehead, John, 86
Whitehead, Richard, 86
Wilbur Smith Bridge, 8, 19, 21-23
Wild, Ebenezer, 140
William L. Clements Library, 259
Williams, Otho Holland, 165
Williamson, Peter, 43
  home of (Battery House), 43
Winds, William, 73

Woodbridge, NJ, 9, 32, 82-83, 90-92
Woodford, William, 57, 84, 121, 221
Works Progress Administration
  (WPA), 317
Writings of Washington, P.
  Angelakos, 258
Wyckoff, Edward H., 24

Y
Yorktown Peninsula, 300
Yorktown, battle of, 96, 122, 214-215,
  217, 300, 302, 316

About the Author

ROBERT MAYERS bases his books on personal site exploration of battlefields, encampments, and places of many critical events of the Revolutionary War. His work sheds light on revered places that have been lost or neglected by history, including sites where patriots fought and died. These are often unmarked, shrouded in mystery, distorted by mythology, and entirely unknown even to local residents. He thrives on discovering facts about the American Revolution not found in the work of earlier writers.

His field trips, when combined with research using original documents and oral accounts of local historians, bring history alive. His readers often comment about how they regret that during their school days that they tuned out history as being distant and dull. His writing can be enjoyed by average readers, not only hardcore military history fans.

Mayers is an active member of several historic societies and is a frequent speaker and contributor to their publications He has given presentations at West Point and the Pentagon and has been featured on Comcast TV and published in the History Channel Magazine, and Garden State Legacy. His service as a combat officer in both the Navy and the Marine Corps provides him with a deep perspective of the many battles depicted in his work.

Bob is a graduate of Rutgers University and has served as an adjunct professor at Seton Hall University Graduate School. As the descendant of patriot soldier Corporal John Allison, the American Revolution has personal meaning for him. He is a longtime resident of Watchung, New Jersey.

Visit him on his website www.revolutionarydetective.com

www.ingramcontent.com/pod-product-compliance
Lightning Source LLC
Chambersburg PA
CBHW020047170426
43199CB00009B/199